Tracking Reason

Tracking Reason

Proof, Consequence, and Truth

Jody Azzouni

OXFORD

UNIVERSITY PRESS

2006

OXFORD
UNIVERSITY PRESS

Oxford University Press, Inc., publishes works that further
Oxford University's objective of excellence
in research, scholarship, and education.

Oxford   New York
Auckland   Cape Town   Dar es Salaam   Hong Kong   Karachi
Kuala Lumpur   Madrid   Melbourne   Mexico City   Nairobi
New Delhi   Shanghai   Taipei   Toronto

With offices in
Argentina   Austria   Brazil   Chile   Czech Republic   France   Greece
Guatemala   Hungary   Italy   Japan   Poland   Portugal   Singapore
South Korea   Switzerland   Thailand   Turkey   Ukraine   Vietnam

Copyright © 2006 by Oxford University Press, Inc.

Published by Oxford University Press, Inc.
198 Madison Avenue, New York, New York 10016

www.oup.com

Oxford is a registered trademark of Oxford University Press

Library of Congress Cataloging-in-Publication Data
Azzouni, Jody.
Tracking reason : proof, consequence, and truth / Jody Azzouni.
p.   cm.
Includes bibliographical references and index.
ISBN-13 978-0-19-518713-7
ISBN 0-19-518713-X
1. Proof theory.   2. Logic, Symbolic and mathematical.   I. Title.
QA9.54.A99 2005
511.3'6—dc22   2005040565

1 3 5 7 9 8 6 4 2

Printed in the United States of America
on acid-free paper

Contents

Part III. Semantics and the Notion of Consequence

Tracking Reason

General Introduction

There are two widespread assumptions that have powerfully shaped philosophy for a very long time. A stab at the first assumption is to call it "the truthmaker assumption": that if a belief, a statement, a sentence—in general, a truth vehicle of any sort—is true, then it's made true (in part) by how the world is in those respects that bear on that truth vehicle. That is, a truth vehicle is made true by what it's about, and how those things it's about are. But this bit of contemporary jargon ("truthmaker") masks how old and venerable the assumption actually is. For almost as long as there has been metaphysics, there have been systematic attempts at recognizing the metaphysical structure of the world by how that world makes truth vehicles true. Contemporary philosophers can, of course, think of the Tractarian Wittgenstein as a particularly rigorous and striking illustration of this sort of program. But the assumption, manifesting in various ways, that something like this strategy is a cogent one has been quite at work well before Wittgenstein—one finds it in Plato, for example, and it continues to powerfully animate philosophy today. Its power is best exemplified, perhaps, in those philosophers who directly *oppose* it; for in opposing it, they overestimate the force of the assumption—how much actually goes if it goes—and use the rejection of it to motivate quite general antirealist (postmodernist, really) doctrines—and now one can think of the later Wittgenstein, although there are many other examples, especially in the wake of the publication of his posthumous work. The thought seems to be that if the structure of truth vehicles is no guide to the limning of the world, then the latter project is simply an incoherent one to begin with.

This is an especially dire development, in my opinion, because contemporary science offers one of the most potentially complete and deep restructurings of our worldview that we've ever faced. Scientific doctrine suggests—if it's true—that the world is shockingly different from what we would have thought (did think) was the case before the scientific juggernaut really got started. And the world is different not only insofar as the *facts* are different, and the laws that govern those facts are different, but insofar as our most fundamental metaphysical assumptions about causation, identity, and so on, are wrong too, and wrong in ways that we can't see our way beyond. (We can't see what the right view of these things is supposed to be.) But unless we make very clear to ourselves exactly how our understanding of the metaphysical structure of the world turns on the truths that science establishes for us, we're likely to both overestimate and underestimate how much science really has changed (or should change) our view of what there is.

The second powerful assumption is one about our ability to reason: that the rules or principles that license the steps by which we reason are ones that, under ideal circumstances, are introspectively accessible to us. The assumption isn't that our minds are so transparent that we see, as it were, the motor of our mind (incessantly turning), that the neurophysiological facts (to put it in a contemporary way) which enable us to reason are visible to us. We recognize (or some of us do, anyway) that such facts are utterly inaccessible to introspection. But we do generally assume, for example, that the logical contours of our concepts (what they entail and don't entail, and, therefore, what follows and doesn't follow from claims we make), although sometimes surprising, are matters that we can ferret out by sophisticated introspection,[1] provided we're careful enough. Logic, for example, isn't taught to our students as a set of principles that we've empirically discovered to govern how we reason. Indeed, the doctrine (held by many) that the laws of logic dictate how we *should* reason (not how we actually—all too often—*do* reason) makes it seem like such rules must be introspectively accessible: Only in this way could we commit ourselves to them to begin with.

What impact this second implicitly held assumption has had (philosophically) depends largely on what sort of scope logic itself is presumed to have. In the modern period, leading up to Kant, certain philosophers—Locke and Hume, most notably—restricted the scope of logic to trivialities, a doctrine some logical positivists, in some moods, tried to resurrect in the twentieth century. Even Quine, well aware of the powerful scope of the first-order subject he demarcated logic within, nevertheless tried to treat it—epistemically—as a chain of obvious moves. One shouldn't rush to characterize "obvious" in terms of a decision procedure—in the case of first-order logic, a decision procedure for proofs. For decision procedures aren't "obvious" in the sense obviously

---

[1] What I've labeled "sophisticated introspection" gives rise to what philosophers call "intuitions," items that are seen as evidentially relevant in, say, formal semantics—but, of course, are seen and used as evidence (if not the final court of appeal) in many other areas (of philosophy) as well.

meant: They need to be discovered, and the recognition of their existence in particular cases is often a profound one that changes the subsequent subject matter. "Obvious," in the sense meant, is that the rules are ones we transparently recognize to be the ones we've submitted our inferences to.

Undercutting the view that our concepts are distinguishable by the introspectable differences they manifest has profound philosophical implications— not all of which I (can) explore in this book. One is a rejection of the visibility of our conventions and norms: Because we take the principles by which we reason to be (in principle) visible to introspection, we tend to enrich "reason" beyond its just deserts. I must be very sketchy here but: (i) We see reason as a tool which enables us to grasp what's possible and what's not, in some broad metaphysical sense—this motivates many metaphysical projects, especially in the metaphysics of modality; (ii) We see the contours of reason—as we grasp it—as marking the limits of rationality: the limits of how it is possible to think coherently at all; (iii) We see reason as indicating, not the bare description of how we do infer results from assumptions we make, but as (grandly put) something much greater: a matter of *norms* (collectively) adopted.

Truth—understood a certain (natural) way—is the centerpiece of both assumptions and is how they knit together. For truth—understood via the truthmaker assumption—points outward toward the world: Truths are about the world, and how such truths are reflects how the world is. And truth, so understood, also points inward as a norm governing reason: Inference is truth-preserving; that is, the mark of a valid inference is that if what it starts with is true, then what it ends with must be true as well. No wonder that those who seek to understand truth in a deflated way so often think that so much goes if (inflated) truth goes: that reason is revealed to be a mere matter of social conformity, and that, as a result, forms of cognitive relativism loom. And (perhaps worse) that the world itself—or any world that *we're* capable of cognizing—is a (mere) construction from within our theories.

My aim, broadly speaking, is to show what does and doesn't follow from a dethroning of the centrality of truth—understood as the truthmaker assumption understands it—from our notions of reason and the world. The rejection of the centrality of truth neither infirms the coherence of metaphysical inquiries, nor brings with it—necessarily—various sorts of anti-realisms, nor leads to cognitive relativism. Rather, what comes into view—once talk of truth is forcibly backgrounded—are the *real* tools we use to structure our understanding of what there is and how we reason about it. Truth is revealed to be a placeholder for philosophically more significant notions.

Showing this, of course, isn't something that can be done within the confines of one book. I did show in *Deflating Existential Consequence* how metaphysics—at least ontology—is still cogent without the truthmaker assumption. No doubt philosophers deprived of that assumption might wonder what's left for them to do. If truth vehicles can be true without there being a metaphysical trace in the world that their truth reflects, how can we go from what we say to what there is? There are other tools available for this purpose. In that book, the focus was specifically on an attempt to establish a genuine metaphysical doctrine—nominalism; but I'm here trying to bring to light the broader methodological moves that were behind that attempt. Similarly, in *Knowledge and Reference in Empirical Science*, I tried to show how realism about theoretical entities was entirely compatible with deflated notions of truth and reference.

I normally like to tackle fundamental, but fairly narrow, issues in philosophy of science and philosophy of mathematics, epistemology, and metaphysics. But the tools I use have much broader scope, and illuminate topics I don't specifically address. Although I'm aware of the broader picture, I don't always do enough to indicate it. In one sense (apart from this introduction), this book will prove no exception: The reader will find a detailed—and sometimes technical—discussion of contemporary work on truth, proof, and consequence. But I've tried (here) to make clear at the outset what the stakes are—in a big-picture sort of way—for the kind of doctrines about truth and reason I'm arguing for.

A word here about the origins of this book. During the period of 1999–2003, I wrote ten or so papers that seemed to fall into two separate groups about two quite distinct topics. One group was in the philosophy of mathematics, on the topic of mathematical proof as it occurs in the vernacular, and a second group presented a new theory of truth that grounds that notion in a generalization of quantification. I subsequently realized that I had systematic and connected views on these topics during 2004, while at work on a pair of papers on *logical consequence* for the Chapel Hill Logic Workshop.[2] My original intention was to publish all the papers in question in a collection. But both referees of the book proposal for Oxford University Press urged conversion of the material into a monograph for easier accessibility to the overall view. I think the resulting book does make the overall view easier to see—in any case, I suppose the papers are otherwise available (if I'm still alive and intact at the point when you're reading this, you can email me, and I'll send you offprints); apart from this, I was able to rewrite the material the way I now see it rather than append clumsy addenda and afterwords to each paper. (My thanks to the referees and to Michael D. Resnik for suggestions about this.)

The result, though, is that some of the papers are only discontinuously present in the current book. Much of the material has been rethought from the ground up. Although the arguments are (usually) the same ones, infelicities and subsequent disagreements with my earlier self have been (often silently) corrected, and the presentation of those arguments, in any case, has often been drastically modified.[3]

[2] At the University of North Carolina at Chapel Hill, April 16–18, 2004.
[3] My thanks to Eric Schliesser for urging me to write an introduction of (roughly) this sort. Although he cannot be held responsible for its contents, I can still blame him for its existence.

I

TRUTH

But what would be a parallel reading of the generalization of 'Tom is mortal or Tom is not mortal'? It would read 'p or not p for all things p of the sort that sentences are names of'. But sentences are not names, and this reading is simply incoherent: it uses 'p' both in positions that call for sentence clauses and in a position that calls for a noun substantive. So, to gain our desired generality, we go up one step and talk about sentences: 'Every sentence of the form "p or not p" is true'.

W. V. Quine (1970, 11–12)

In 1968, then, there emerged footnote 3 of Kaplan's "Quantifying In": if the quantifyings in are meaningless, why not assign them meanings ... ? Splendid!

W. V. Quine (1986, 291)

INTRODUCTION TO PART I

Truth is, and has always been, a central topic in philosophy. Direct interest in the word *true* itself may ebb and flow, but it's never far from center stage. At present it's a particularly popular word to write about, if only because the many wide-ranging positions that philosophers currently have on truth allow so many *other* topics from philosophy to come clearly into range. To a large extent, any major rethinking in philosophy requires both destruction and construction. One must circumvent, undercut, and directly challenge alternatives—both real and imagined—and one must nevertheless leave enough room in one's exposition to present the new view. I engage, therefore, with the rich contemporary literature on truth judiciously, especially because my aim in this book isn't only to engage with the notion of *truth*, but also to engage with two notions taken to be closely related to truth: *consequence* and *proof*.

I am, broadly speaking, a minimalist. These days there are so many species of minimalist, however, that to claim to be a minimalist isn't to be very informative about one's position. Minimalists—this much is true of all of them, I think—are very much "naysayers." One or another "substantialist" notion of truth, e.g., "truth as correspondence," has found itself playing many roles in the hands of philosophers: as central to ontology, e.g., the commitment to objects via the "truthmaking" requirement on truths; as central to a theory of understanding of a language, via the grasping of "truth conditions"; as central to epistemology, as the rationale for epistemic practices that are seen as "truth-seeking"; and as central to a theory of inference, taken here as a theory of "truth-preserving" moves in a language. And these hardly exhaust the roles that this overwhelmed and sadly fatigued idiom has been conscripted for.

Minimalists deny that "truth"—the purported substantial notion, anyway—serves these roles; often they claim that "true" plays a humble expressive role that facilitates communication, and that's *all* it does. But (some) minimalists have gone much further. If they can't show that their minimalist notion of truth can serve the same role that the "substantial" notion of truth would have served, in ontology, or in a theory of understanding, and so on, they draw, as a result, very dramatic philosophical conclusions about these other topics. They have been known to adopt, for example, Wittgensteinian-flavored views about understanding and inference, deflationist views about properties, and relativist and irrealist views about, well, *truth*. The evaporation of a "substantial" notion of truth, that is, is often taken to be accompanied by the evaporation of other substantialist doctrines about understanding, ontology, and epistemology.[1]

I'm definitely *not* a minimalist in this sense. That "true" *doesn't* play a certain role in ontology, epistemology, and so on, doesn't mean *nothing* does. One reason, although not the only one, for including analyses of the notions of *consequence* and *proof* in this book is to show that robust construals of these notions are still available even without the backbone of a substantial truth idiom. Proof, for example, needn't degenerate into a socially constructed ghost, a matter of mere social agreement, just because a substantial notion of truth isn't available to fix what it is that proof supposedly tracks. The impression to the contrary is due to the history of the field: that various kinds of antirealists, social constructivists, and so on, have thought that the fastest way to establish their broad metaphysical positions is to, as it were, chop down TRUTH—to show that one or another substantial notion of truth is false. But this is a wrong-headed strategy: Substantial truth *can* go—but that doesn't mean that the various sorts of realism *must* go with it.

So, as I indicated in the general introduction, I see the job of this book—which given its magnitude can only be partially undertaken here—as exposing the elements that actually undergird fundamental notions in ontology, epistemology, and philosophy of language, elements which move into clear view once it's recognized that the idiom of *true* isn't the backbone of these notions. Realism—at least my version of Realism—doesn't need a substantial notion of truth to be robust.

There is at least one other aspect of the notion of truth that has bedeviled much of the literature, which I attempt to straighten out. This is the somewhat technical question of what sort of idiom "true" is. It appears—both in the vernacular and in the formalization of that notion at the hands of Tarski—to be a predicate, and one that's codified or even defined in terms of T-biconditionals, statements of the form: *"S" is true iff S*. In a way that will be made clear, this leads pretty directly to an "immanent" notion of "true": one that applies directly only to one's own language—or worse, idiolect—and that can only derivatively, by translation say, be applied to other languages. I show

---

[1] Perhaps the purest example of this sort of tendency can be found in Rorty 1991. Also see Horwich 1998.

that a characterization of *true* that treats it as a species of quantification escapes the problem of immanence.

Marian David (1994, 52–60) describes a number of motivations that many philosophers have for being deflationists or disquotationalists about truth.[2] Among these are animosities toward abstracta and proclivities toward physicalism. Although, as my books *Knowledge and Reference in Empirical Science* and *Deflating Existential Consequence* make clear, I share both these animosities and proclivities—I'm a nominalist, and my ontic commitments fall into the physicalist camp (broadly speaking)—these motivations have little to do with why I'm the sort of deflationist I've become. One reason for my deflationism is that I see deflationism about the role of "true" in the vernacular (that it functions as a logical device to facilitate semantic ascent and descent) as entirely compatible with even a rich correspondence view of truth that takes sentences (say) to correspond to structured facts containing objects, properties, and whatnot. My deflationism about the role of "true" is driven—as far as I can see—by evidence of usage, and by nothing ontic whatsoever.

As it turns out, I'm also opposed to a rich correspondence view of truth—but here too, the reasons lie apart from ontic inclinations. For most philosophers, austere ontic inclinations require their rewriting various theories so that whatever they abjure (ontically) not be quantified over in the resulting theories (to use the Quinean locution). But since I don't see the ontic commitments of a theory as captured by what it quantifies over (by what it's, generally speaking, "about") I'm free to use all sorts of mathematical locutions in theories; this includes semantic theories. So it's perfectly acceptable (to me) if the semantics of natural language contain a rich texture of *talk* of correspondence relations. But I mustn't give the appearance of genial agreement with the traditional correspondence theorist by saying this—for that theorist has metaphysical aspirations. By virtue of the rich correspondence relations he takes himself as committed to, he also takes himself as committed to the objects apparently in such relations. " 'Snow is white' is true," commits him so he thinks to snow and whiteness, and (for some) *even* the fact these reside in. So too for "$2 + 2 = 4$." Our disagreement, therefore, is at root a metaphysical one over what has come to be called (in the literature) the "truthmaker" assumption: that truths must have relata by which they are made true. I deny this, but, as I said, my denial doesn't necessarily betray a disagreement about the semantics of languages.

I'll conclude this introduction with a brief description of each chapter in part I. Chapter 1 sets out the contrast between metaphysical truth deflationists and biconditional truth deflationists: those, roughly, who are deflationist about the truth idiom—what it's used for—and those who are deflationist

---

[2] There have been attempts to distinguish the meanings of these terms in the literature: deflationism, minimalism, disquotationalism. I'll (more or less) use these terms indistinguishably, and instead indicate differences in positions by directly describing them. In some sense, as the reader will see, I'm all three of these things. In another sense, I'm not. The details of the position—what's accepted and what isn't—must be identified by the theses I accept and deny: The nomenclature in this area operates too much at cross purposes to be of value.

about the possibility of saying something metaphysically substantial about truth. The chapter also motivates the deflationist view of the truth idiom: that "true" is used in the vernacular only to facilitate blind truth-endorsement—even when giving truth conditions.

Chapter 2 is dedicated to showing that the vernacular truth predicate facilitates blind truth-endorsement of statements we neither understand nor know that a translation exists for. This "transcendental" truth predicate eludes formalization Tarski-style or, for that matter, by sentential substitutional quantification. Chapter 2 establishes this, and motivates the presentation of a formal tool, *anaphorically unrestricted quantifiers*, that can successfully execute the apparent transcendental role of the vernacular truth predicate.

Chapter 3 introduces the reader to a formalization of anaphorically unrestricted quantifiers. Their model theory and proof theory are given, and it's shown how they enable blind truth-endorsement of statements not in the language of those quantifiers. This is the most technical chapter in the book. The reader disinclined toward such can skim it for philosophical content, and move on.

Chapter 4 turns to the question of the relationship of anaphorically unrestricted quantification (which is a formalization) to the notion of truth in ordinary languages. It's argued that the appropriate relation is one of regimentation; I spell out in some detail what I mean by this. In particular, it's argued that regimentation isn't the building of a new linguistic home for those brave souls willing to desert the vernacular. Instead, it's a normative guide to inference for practitioners remaining at home.

Chapter 5 provides further motivation for regimentation by indicating the reasons for thinking that ordinary languages are inconsistent in just the way that Tarski thought they were. It's shown, nevertheless, how the reasoning of speakers of the vernacular (that is, all of us) can be recognized to be coherent, and how semantics can be provided for that reasoning—despite the inconsistency—by regimentation.

Part I is constructed on the basis of materials drawn, in part, from the four papers listed below, and from a talk I gave at the logic workshop at the University of North Carolina at Chapel Hill on April 16, 2004.

(1) Truth via anaphorically unrestricted quantifiers. *Journal of Philosophical Logic* 30: 329–54, 2001.

(2) The strengthened liar, the expressive strength of natural languages, and regimentation. *Philosophical Forum* 34 (3 and 4): 329–50, 2003.

(3) Anaphorically unrestricted quantifiers and paradoxes. Forthcoming in *Deflationism and paradox*, ed. Brad Armour-Garb and JC Beall. Oxford: Oxford University Press.

(4) Tarski, Quine, and the transcendence of the vernacular "true." *Synthese* 142: 273–88, 2004.

Apart from my continuing thanks to those acknowledged in the above papers for their help, I also wish to thank Douglas Patterson for looking over a version of part I and for sending me a number of very useful comments and criticisms.

1

Truth and Truth Conditions

1.1 Biconditional Truth Deflationists and Metaphysical Truth Deflationists

Biconditional truth deflationists typically see *truth* as a *predicate* to be directly applied to sentences or to propositions (and, for some of them, what it's *directly* applied to—sentences or propositions—matters a great deal); but, in any case, it's taken to be a predicate governed by the infinite set of (nonpathological) T-biconditionals resulting from every nonpathological sentence of a language (typically, of the deflationist's own language)—or every proposition—appearing in an instance of the schema: *"S" is true iff S.*[1] Some deflationists say that "claims like **It is true that S** and **The proposition that S is true** are trivially equivalent to S, and this equivalence is in some sense definitional of the notion of truth."[2] (Sometimes) accompanying this view is a still grander philosophical claim to the

---

[1] It's highly nontrivial, of course, how to separate nonpathological T-biconditionals from pathological ones—such as liar paradoxes—since, in general, the latter (as a class) aren't syntactically distinguishable. Indeed, another related problem facing the biconditional truth deflationist (because the truth predicate itself can appear on the right side of such biconditionals), is that it's highly nontrivial exactly how, on such a view, the crucial T-biconditionals are supposed to be systematically generated (see McGee 1992). It won't do, therefore, to simply postulate a "disposition" to assent to such things when presented with them. In chapter 4, I make clear why my version of deflationism doesn't face these problems.

Here, and throughout the book, where quasi-quotes or other use/mention smoothing devices are required, they're implicitly understood as present and correctly operating.

[2] Soames 1997, 4.

effect that the truth predicate doesn't pick out a property or (perhaps) a *real* property; that, anyway, talk of truth shouldn't appear in certain explanations the way it ordinarily seems to. This last bit is cashed out—by some—as a program to (i) scan through the various uses of "true," when we give truth conditions in semantics, for example, or when we give various explanations about certain intentional states, such as beliefs, desires, and so on, and to (ii) either eliminate such usages of the truth predicate, or otherwise show their compatibility with the deflationist view. Because, and now I'm engaging in diagnosis, the biconditional truth deflationist takes the conceptual status of the idiom of truth to essentially be predicational, attempts to show that (certain) explanations needn't involve talk of truth often look similar to various eliminativist-style programs typically directed toward classes of predicates seen (by some philosophers) as similarly unacceptable in explanations. These programs try to show that such classes of predicates can be rendered explanatorily idle via alternative explanatory statements that can do the same job but from which the undesirable predicates are absent.

Deflationists—recent ones, anyway—when pressed about the *point* of the truth predicate, give a reason first found in Quine 1970: that the purpose of the truth predicate, *when coupled with a suitable quantifier*, is to enable maneuverability between use and mention in the context of blind truth-endorsement.[3] One often needs to single out classes of statements that one can't directly assert or deny (by *using* them, that is, or by using negations of them)—because they are too numerous or because one doesn't know what they are, or because one doesn't know how to say them—and yet one must still somehow manage something that, for the purposes at hand, is *as good as* using them. This minor miracle of articulation is achieved by coupling an ordinary *description* of the specific sentences needed (to pick them out) with the use of *another* predicate that *describes them* so that *the same purpose as* using those specific sentences is served.[4] I stress again: The truth predicate is, as Quine puts it, "a device of disquotation." It's not *itself* a device of

---

[3] By the *purpose* of the truth predicate, I mean this: a purpose that, given the resources of natural languages, isn't satisfiable *without* using the truth predicate. Like any item in natural language, truth predicates are put to *many* uses—ordinary language is always flagrantly opportunistic with respect to its available resources (misleadingly so, at least when it comes to innocent philosophers); but those other uses (e.g., "that's true," when said immediately after a remark one wants to assent to, or, " 'snow is white' is true") are ones easily facilitated (except among the *very* lazy) without the truth predicate. Not so of its use in (many) blind truth-endorsements.

By "blind truth-endorsement" I mean uses of the truth predicate where the sentences endorsed or denied don't themselves appear, e.g., "What John said is true." In previous work, I described these as "blind truth-ascriptions," but Douglas Patterson has urged that, in the current polemical deflationist vs. inflationist atmosphere, this nomenclature may be misleading. He has suggested "blind truth-endorsement," and I've embraced it (along with a coined complementary "blind denial" or "blind false-endorsement"). I'll often, when speaking generally, describe both blind denials and blind endorsements by "blind truth-endorsement" or "blind endorsement."

[4] I've borrowed Quine's nomenclature: use/mention, but what's being described is actually better labeled as a use/description distinction.

generalization: a quantifier (of some sort); rather, it's something much more humble that's needed, typically, when quantifiers are available that range over sentences (or propositions)—when such quantifiers, that is (by virtue of what they range over), involve semantic assent: a switch from talking about whatever it is that one's sentences enable one to talk about to talk about the sentences themselves.[5]

Deflationists from the middle of the last century, or philosophers, anyway, then sympathetic to a deflationist construal of "true," seemed strangely unaware that the indispensability of the truth idiom is due only to the practical need for the expression of ineliminable blind truth-endorsements.[6] One finds writers, otherwise notoriously sensitive to the nuances of the vernacular—such as J. L. Austin—making careful note of how "true" is used in English, what objects (e.g., beliefs, propositions, sentences, assertions, etc.) are described as true, how it's used to indicate agreement ("That's true"), and so on. What's surprising is that all the uses of "true" that such philosophers evince awareness of, even when blind truth-endorsements, seem eliminable ones: uses we could do without merely by added circumlocution. This is explicit, for example, in Austin 1979 (118) where he offers as illustrations of "the primary forms of expression" of saying something is true: "It is true (to say) that the cat is on the mat," "That statement (of his, &c.) is true," and "The statement that the cat is on the mat is true." Although examples of blind truth-endorsements are mentioned explicitly by Frege, Tarski, Austin, Strawson, and many others, they aren't seen as central to the illumination of the truth idiom.[7] One searches in vain (prior to Quine 1970, anyway) for *any* indication of awareness that it's blind truth-endorsement, in particular, *ineliminable* blind truth-endorsement, that is the *raison d'être* for the presence of the truth idiom in

[5] Contrast this with Horwich (1997, 96) where "the *value* of our concept of truth" is described as "its utility as a device of generalization." Also Field 2001b, 153: "There is nothing in deflationism that prevents the use of 'true' in explanations as long as its only role there is as a device of generalization." I agree, as the forthcoming indicates, provided the phrase "device of generalization" is corrected to "device of disquotation," and provided that deflationism—understood as a claim about the role of a piece of language, "true"—is distinguished carefully from other doctrines (as I'll endeavor to do later in this chapter).

Of course "true" isn't *literally* a device of disquotation or semantic descent *either*, although calling it *that* is far closer to the *truth* than describing it as a device of generalization. But a *real* device of disquotation or semantic descent would be a quotation-*canceling* device. The intuitive recognition that T-biconditionals are true or even "necessarily true" is insufficient to establish a thesis about the truth predicate that strong! (Otherwise "Everything John said is true" would seem grammatically ill-formed.) To put the point precisely (again) "true" is a predicate that we take to hold of sentences that we would use (rather than describe) *if we only could*. Nevertheless, for ease of exposition, and in order to remain comfortably within the nomenclatural tradition, I'll continue to call "true" a device of semantic ascent and descent or a device of disquotation.

[6] See chapters 1 and 2 of Azzouni 2004a for extensive illustrations of the role of ineliminable blind truth-endorsements in ordinary life and in the sciences. I reprise this role, briefly, in 2.2.

[7] Strawson (1999, 176) writes: "Sometimes, to embarrass, or test, our audience, we use . . . 'What John said yesterday is true.'"

the vernacular—that is, that it's because of such endorsements, and them alone, that ordinary language requires a truth idiom.[8]

In any case, describing the role of the truth predicate as essentially an enabler for blind truth-endorsements makes the corresponding role of T-biconditionals, *"S" is true iff S*, transparent. The purpose of this list is *not* to indicate that attributions of truth to sentences or propositions are analytically equivalent to, or a priori recognizable as, the sentences or propositions themselves; it's only to fix the extension of the predicate "true" so that, when coupled with a suitable quantifier, it can function as needed. I should add that it's a very good thing that the deflationist needn't join with the redundancy theorist on the claim that the right and left wings of the T-biconditionals mean the same thing, because even a glance reveals that they *don't!* The subject matters of the left sides of T-biconditionals are either sentences (or propositions, depending on the view) and their truth; the subject matters of the right sides of T-biconditionals are *varia*: the richly infinite spread of things (that exist and that *don't*) that we talk about.[9]

Although (at the end of the day) I'm not much of a fan of either analyticity or aprioricity, it may help to understand the import of T-biconditionals by distinguishing *two ways* that *any* biconditional can be understood as either analytic or a priori. The first is if one wing is analytically (or a priori) equivalent to the other. "A bachelor is tall iff an unmarried male is tall," on one meaning of "bachelor," fits the bill here. But another way is if the biconditional itself is deducible from definitions or principles themselves a priori or analytic. Then the biconditional is analytic (or a priori) not because each wing is analytically or a priori equivalent to the other, but only because the biconditional *as a whole* is deducible from something else and *that fact* makes it analytic or a priori. T-biconditionals are, at best, analytic, trivial, or a priori in this second sense, and so it isn't natural to suggest that one wing of a T-biconditional

[8] Thus, *my* biconditional truth deflationist is *not* correctly characterized by Gupta (1999, 287) when he writes that for deflationists "the disquotation thesis is understood . . . as saying not just that the T-biconditionals are true, nor just that they are necessarily true. The claim is rather that the T-biconditionals issue from our very understanding of 'true', that they explain (at least partially) the meaning of 'true'." My biconditional truth deflationist accepts the centrality of the Quinean insight about blind truth-endorsement—and reads the meaning of "true" off of its functional role. Gupta's characterization, however, does fit many, if not most, contemporary deflationists, as well as Austin, Frege, and others.

[9] As Soames (1997, 43) characterizes (a version of) the redundancy theory, it attempts to escape this obvious fact by distinguishing grammatical from logical form, so that, for example, "*The proposition that snow is white is true* has the same simple logical form as the sentence *Snow is white*." Apart from a flagrant disregard of the linguistic facts, this move endangers the utility of the truth predicate in blind truth-endorsements. See Horwich 1998, 39 n. 20, for a list of citations of objections to the redundancy theory along these lines. I'm unsure how many deflationists, apart from redundancy theorists and (some) prosententialists, actually commit themselves to an identification of the meaning or content of the two wings of T-biconditionals; Patterson (forthcoming(a), n. 16) attributes such an identification to Quine (1970), Leeds (1978), and Field (2001a), among others; but I've reservations about the first two. In any case the point is that, as far as the role of "true" is concerned, biconditional truth deflationists *needn't* so commit themselves.

means the same as the other. I can go further. Consider a sentence of one's language that one doesn't know the meaning of. The T-biconditional of that sentence is one that we grasp the truth of even though we don't grasp the meaning of either wing of it. We deduce the equivalence of the wings of a T-biconditional via our understanding of "true" *and* quotation.

It may seem like nit-picking to stress that "true" isn't a device of generalization, but rather a predicate that enables the job of asserting collections of sentences without actually using them. I think not, since being clear about this can recast certain philosophical issues. I mentioned earlier that some deflationists think that "true" doesn't pick out a property, or at least not a real property. Apart from an intrinsic unclarity, due directly to what the notions "property" and "real property" are supposed to mean, being precise about the truth predicate's actual role distinguishes rash biconditional truth deflationists from careful biconditional truth deflationists. Here's how: One can couple the truth predicate with ordinary quantifiers that range over (among other things) sentences or propositions. In that case, "true" is exactly like any other predicate insofar as it singles out a subset of the domain of sentences or propositions, just as "begins with the letter 'r' " does (with respect to a domain of sentences). The point that "true" coupled with other devices enables us to assert collections of sentences, and the related point (see 1.3) often made in the literature that if we could assert infinite disjunctions and conjunctions, "we wouldn't need 'true' " are both irrelevant to the fact that, nevertheless, "true" as described here is an ordinary predicate.

One can nevertheless legitimately *worry* about this "ordinary" predicate "true," be concerned that its extension must be fixed in such an odd way (by T-biconditionals); related to this, one may worry that truth doesn't seem to be definable in terms of other more acceptable ("physicalistic," say, or "nonsemantic") predicates; but these issues go well beyond what the biconditional truth deflationist can carefully claim on the basis of the role or meaning of the truth predicate alone: The truth predicate has a certain logical role (but, despite that, it's still a *predicate,* with an extension like any other predicate). This role, that "true" has, *is compatible with* a claim commonly made by a more radical kind of deflationist: that in fact the propositions that "true" applies to have nothing in common, no common nature, and so (in this admittedly still vague sense) "true" doesn't pick out a real property. I'm sympathetic with a certain version of this view, and I'll give some considerations in its favor shortly.[10] But all I want to stress now is that no argument

---

[10] Horwich (1997, 99) claims that truth "doesn't have [an underlying nature], indeed *couldn't* have one." He then refers to an argument he gives in his 1995 article. Although the argument there doesn't concern *truth,* but rather, reference, it's easy to see why he thinks (a variant of) it applies here: "The notion of reference, insofar as it satisfies the disquotational principle, enables [the capture of] certain generalizations that cannot be captured merely by using the usual devices (that is, 'all', 'every', or the universal objectual quantifier). If this is right, then facts articulated with the concept of reference (including those that correlate reference with other properties) cannot be deduced from ordinary non-semantic generalizations and, therefore, cannot be explained by them" (78). But explanations of *P* are relative to descriptions of *P*; what's

for it can be based on what the role of the truth predicate is, nor on the claim—even if sustained—that this role doesn't require a *predicate*. The role of the truth predicate is neutral on whether "true"—the predicate—picks out something substantial or not: whether "true" is coextensive with, e.g., a non-semantic predicate.[11] As I show in 1.8, arguments about this issue will be to the point only if they focus on purported uniformities among *truths*, rather than on the role of "true." This suggests that we should refine our categorizations of truth deflationists. There are biconditional truth deflationists, as described in the opening paragraphs of this chapter: They are concerned with the predicate "true." But there are also those who claim that the sentences or propositions that "true" applies to have no common nature. Let's call these *metaphysical truth deflationists*. I'll have more to say about the latter sort of deflationist later; suffice it to say now that BTD (biconditional truth deflationism) and MTD (metaphysical truth deflationism) are doctrines that look entirely independent of each other; I mean, that is, that it looks as if one can be committed to one or the other but needn't be committed, as a result, to both.

One last point is crucial (especially to what follows). I perhaps overly deflated what the careful deflationist is committed to when I described her as claiming that "true" is a predicate needed only for semantic ascent and descent—when coupled with certain quantifiers—and that such a view is compatible with "true" as a predicate either being coextensive with something substantial or not (with, that is, the truth or falsity of MTD). This compatibility claim would be violated if a use of the truth predicate were found, in ordinary parlance, that was *not* a mere matter of semantic ascent and descent but presupposed coextensiveness with some other (nonsemantic, say, or

---

deducible from what is relative to the description of the second "what." So (i) a truth predicate could be coextensive with another nonsemantic predicate **R**; (ii) although one might not be able to deduce (all the) "facts articulated with the concept of truth" from **R**, that wouldn't fault the attribution by **R** of an underlying nature to truth. (Moral: it's very risky to make metaphysical pronouncements—say, about the underlying nature of things—on the basis of what can be explained by what.) See Gupta 2002 and Patterson forthcoming (b), both of which put to interesting uses the possibility of a characterization of the extension of the truth predicate from which T-biconditionals aren't derivable.

[11] This neutrality view, of course, is held by a number of philosophers, even when, otherwise, their construals of "true" vary greatly. For example, both Soames (1997) and Lance (1997) endorse the claim. Arguably, so does Tarski (1944). On the other hand, many philosophers *contrast* "deflationism" with substantial theories of truth—see them in conflict. Horwich (1998) takes this view; so does Gupta (1999). David's (1994) book is entirely structured on the assumption of such a conflict between the two families of views.

One might try to undercut the neutrality view by arguing this way: Even granting that the truth idiom is only for the purpose of (helping) facilitate blind endorsements, in order for it to successfully implement this role, its semantics requires a substantial construal of the property of truth. I'll indicate one place (in 2.7) where such an argument might arise. I should add, in any case, that I've a number of other independent reasons against the move of interpreting the truth predicate in this (substantial) way—see 1.8.

physicalistic) predicate.[12] The careful BTDist must make sure that no such presupposition arises when the word "true" is used in the vernacular.

1.2 Uses of "True" in the Vernacular

So let's consider ordinary and natural uses of the truth predicate in various sorts of explanations and statements. It's perhaps appropriate to first give several examples (not all of which will be discussed in this chapter, or even in this book) to illustrate the range of cases the BTDist must consider:

(A) "Snow is white," is true iff snow is white.

(B) For all sentences A and B, $(A \& B)$ is true if and only if A is true and B is true.

(C) A sentence S_1 implies another sentence S_2 iff: if S_1 is true, then S_2 *must* be true as well.

(D) The theory T is true, and E is an implication of T, so E is true also.

(E) John said, "snow is white," and everything John says is true, so snow is white.

(F) Peter is as successful as he is, generally, because most of his beliefs are true.

As I understand the BTDist, she takes herself as required to show that the uses of "true" as they appear in examples (A)–(F) are all disquotational uses, or she must replace such examples with ones serving the same role, but in which the truth predicate, if it appears at all, does so only as a purely disquotational device.[13] Call this *the project of deflationist exegesis.*

---

[12] Williams seems to have the same point in mind. He writes (2002, 153): "In so far as we deflationists distinguish between predicates that 'stand for substantive properties' and those that do not, it is in terms of the use that we find for the predicates in question. Where we find an indispensable explanatory use, we recognize a substantive property. If all we find is, say, a device for semantic ascent, we do not." I would qualify this remark in two ways. First, if the indispensable explanatory use of a truth predicate *is* as a device for semantic ascent, the deflationist won't be disturbed. (And here, by "indispensable explanatory use" I mean only "its ineliminable use in an explanation.") By way of my second qualification, I would direct the reader to the second paragraph of note 11.

[13] So there is in play (at times) a project of replacing truth conditions with something else altogether. I'll use Field 2001a as an example. Field writes (108): "The main idea behind deflationism . . . requires only that what plays a central role in meaning and content not include truth conditions (or relations to propositions, where propositions are conceived as truth conditions)." Also on page 108: "If deflationism is to be at all interesting, it must claim not merely that what plays a central role in meaning and content not include truth conditions *under that description*, but that it not include *anything that could plausibly constitute a reduction* of truth conditions to other more physicalistic terms." On the other hand, we find (106–7): "Even a crude verificationist can grant the legitimacy of talk of truth conditions of his own utterances. . . . A pure disquotational notion of truth gives rise to a purely disquotational way of talking about truth conditions." The contradictory feel of these quotes can be dispersed, I think, if we recognize that Field thinks that deflationist truth can sometimes play a role in statements wherein we ordinarily use the word "true," and that sometimes

(E) and (F) are cases in which it's often thought that "true" plays a crucial explanatory role, one that the anemic notion of "true" that BTDists allow themselves can't manage.[14] (C) and (D) are examples of sentences that illustrate how, intuitively, talk of truth seems bound up with our notion of implication: We take implication to be "truth-preserving"; indeed, it may seem that the appropriate way to *define* a valid inference is as one that "preserves" truth. (C) illustrates this idea. (D) seems to provide an example of how we use the truth-preserving properties of implication to infer truths from other truths. I'll take a closer look at what's going on with (C), (D), and their ilk in part III of this book.

For now, let's focus on (A) and (B). (A) seems to be one of those trivial uses of "true," that have (historically—recall 1.1, especially note 9) given philosophers the impression that redundancy theories of truth are plausible: that "true" has only an eliminable role in our language. (B) has a similar tautological flavor, given, of course, that we understand the word "and." Perhaps surprisingly (for the novice in philosophy of language, anyway), statements of these sorts aren't taken to be trivial matters at all: They are examples, indeed, paradigmatic examples, of the giving of "truth conditions" for statements.[15] Such clauses are taken by many philosophers as crucial to a theory of meaning of a language or as crucial for a theory of the understanding of a language (or both).[16] One might try to intuitively motivate this sort of program like so: One knows what a sentence

---

it can't. Indeed, he says (108) that one purpose of his argument is to make explicit "the limited role that I think the deflationist can give to truth conditions, and identifying kinds of role that deflationism cannot allow truth conditions to have." I'll argue later in this chapter that BTD places no constraints on the role of the truth predicate *at least for the uses of "true" in the contexts of the giving of truth conditions.*

Notice there are two separable issues here for the BTDist. The first is whether, as things stand, a nondisquotational truth predicate occurs in the vernacular. The second is whether, if nondisquotational uses of the truth predicate are found, they are *indispensable* to the language.

[14] The problem that some philosophers have faced (see, e.g., Field 2001b, 151–52), when the truth idiom interpenetrates belief ascription, is that an immanent notion of "true," one that weds it to the T-biconditionals of sentences of the language of the speaker, infirms use of it when the individual to whom the belief ascriptions are made doesn't speak the same language as the belief-ascriber. One must then invoke one or another notion of translation that, in its train, seems to introduce tools that the truth deflationist isn't entitled to. It's at this point I part ways with the BTDist. See chapters 2 and 3.

[15] Truth conditions, thus understood, arise in the seminal Tarski 1932 publication (1983a). There, in the course of his axiomatizing and defining a formal notion of truth, he gives what has come to be described as a "compositional semantics" for the languages he investigates. Such a semantics requires, among other things, clauses of the form (B) for the various logical idioms of the language, and what amounts to clauses of the form (A) for sentences without logical particles. The approach has proved widely influential and has been successfully applied to languages with idioms, modal ones, for example, that Tarski never considered. There are numerous expositions and technical variations of his original recursive characterization of the truths of a language via clauses in the spirit of (A) and (B). See, e.g., Quine 1970 or Soames 1999. For a controversial discussion of what Tarski did and didn't achieve by means of his semantic theory of truth, see Field 2001c.

[16] Again, the approach—as generally described—is extremely influential. Perhaps the *locus classicus* for it occurs in early papers of Davidson, e.g., 1984c, 1984d, 1984e. But there are many others.

means ("understands" a sentence) if one knows the conditions under which it's true and false (its "truth conditions").[17] Perhaps this sounds like rather a lot to have to know to understand a sentence or to grasp its meaning—(Am I supposed to know *all* the conditions under which a sentence can enjoy truth and falsity? If so, *know* in what sense? Have the capacity to *distinguish* such circumstances?)— but then it's pointed out that knowing truth conditions isn't so hard after all: Knowing that "Snow is white," is true iff snow is white *suffices* because the right side of this truth-condition-giving clause really does give (all) the conditions under which "Snow is white," is true. But (the worry then becomes) if that's *all* that truth conditions come to, they look awfully trivial to be central to a theory of meaning or of understanding. There are responses to this conundrum, of course;[18] but my purpose at the moment isn't to explore the question of whether truth conditions thus understood can be used in a successful theory of the meaning of the sentences of a language or in a successful theory of the understanding of a language, but to explore whether truth-conditional analyses as encapsulated in (A) and (B) are uses of the truth predicate inaccessible to the BTDist.

1.3 Deflationist Exegesis

Let's return, therefore, to the BTDist project of deflationist exegesis, with respect to (A), (B), and their ilk. One way to characterize the role that the BTDist attributes to the uses of "true" in the vernacular is to note, as Putnam 1978 (15) does—following Leeds 1978—that "if we had a meta-language with *infinite conjunctions* and *infinite disjunctions*... [and if] we wanted to say 'what he said was true'... we could say instead:

(1) [He said 'P_1' & P_1] or [He said 'P_2' & P_2] or... "

That is, we wouldn't need the word "true" at all. But we can't use infinite disjunctions and conjunctions. "So," (Putnam 1978, 15) "we look for a finite expression equivalent to (1)":

---

[17] Describing the truth-conditional "tradition" as one "whose early advocates include Frege, Russell, the early Wittgenstein, and Ramsey," Field (2001a, 104) writes that a "strong prima-facie reason for the attractiveness of this position is that the way we standardly ascribe meanings and contents is via 'that' clauses, and the ascription of 'that' clauses is in effect the ascription of truth conditions: to describe an utterance as meaning that snow is white, or a belief state as a state of believing that snow is white, is in effect to say that the utterance or belief state has the truth conditions that snow is white." He continues: "Since 'that' clauses and hence truth conditions play such a central role in our *ascriptions* of meaning and content, it would seem as if they ought to play a central role in the *theory* of meaning and content," although he adds that it isn't easy to see precisely what that central role is supposed to be.

[18] For example, Davidson 1984a, 1984c, and 1984d focus on the constraints that he takes to arise from the *whole* Tarskian theory that's to be applied to a natural language, and from which the T-biconditionals emerge as theorems: finite axiomatizability (which is a theoretical constraint on a theory of the meaning of the sentences of a language corresponding to a learnability requirement), recursive characterizations of the logical particles of the language, extensionality, etc., which jointly prove nontrivial in application to natural languages, and indeed, to many formal languages. I'll say more about this issue later (1.6).

(2) For some x he said x & x is true

is equivalent to (1) provided for each i ($i = 1, 2, 3, \ldots$)

(3) "P_i" is true if and only if P_i.

Putnam's argument, presumably, isn't supposed to establish that the truth predicate is the *only* finitary device that can substitute in blind truth-endorsements for infinite conjunctions and disjunctions; that clearly isn't the case. In moving from (1) to (2), one particular syntactic decision among other options has been enacted: As "[He said 'P_1' & P_1] ..." stands, we have quote-names of sentences *and* the sentences themselves. Reworking (1) into something finite via quantifiers calls for anaphora—the referential connection of pronouns (or variables playing their role in a formalism) back to a quantifier—but this can be done in one of *three* ways. One can turn all the occurrences of "P_1," "P_2,"..., into nominal contexts: This is Putnam's route, which calls for a predicate ("true") and names to appear in those contexts where "P_1," "P_2," ..., stand alone; that is, it involves nominalizing apparent sentential contexts. *Or*, one can treat all occurrences of "P_1," "P_2,"..., as occurring in sentential contexts (including the context, "He said 'P_1'"): This is the strategy of Grover et al. 1992, which takes nominal contexts, what otherwise look like predicates followed by quote-names of sentences, as composed of sentences and sentential operators instead; last, we can leave the apparent presence of both nominal and sentential contexts untouched by quantifying into those contexts *simultaneously*.

Why *do* natural languages handle truth—at least as far as surface syntax is concerned—predicationally rather than sententially? Soames (1999, 34) notes that, in *English*, although we can generalize from "John's mom said that John solved the problem," or "Bill's mom said that Bill solved the problem," or "Harry's mom said that Harry solved the problem," and so on, to "Some man's mom said that he solved the problem," we can't generalize from "$1 = 1$ although no one can prove $1 = 1$," or "$1 = 2$ although no one can prove $1 = 2$," and so on, with every sentence, to "Some sentence although no one can prove it." We have to say: Some sentence is true, although no one can prove it's true.[19]

But *why?* Soames writes (1999, 34): "Since quantifier phrases like these are noun phrases in English, neither they nor the pronouns they bind can occupy the position of sentences in English." Two points, I think, are worth making. First, Soames's remark involves an implicit (and easily overlooked) constraint. It's *logically possible* for quantifiers to simultaneously bind pro-sentences *as well as* pronouns. What stops this in natural languages is that, in such languages, anaphora won't cross syntactic borders. Much of the complexity of "true" (indeed, I dare say, that a *predicate* "true" in English exists *at all*) is because of this constraint; an example of the complexity that arises when a predicate "T" is introduced, for example, is that we can't express a

---

[19] I take it that other natural languages are analogous on this point.

T-biconditional *generalization* to govern it like so: "$(p)(Tp \Leftrightarrow p)$."[20] Proponents of prosentential truth also accept the anaphoric constraint; that's why they insist on transforming prima facie nominal contexts into sentential ones.

Second: Soames's considerations (which I'm otherwise swayed by) make the presence of a truth *predicate* in English a parochial fact about *English* (and other natural languages). Because quantifier phrases are noun phrases, we need a predicate that employs pronouns to generalize over sentences. One can ask *why* natural languages are so structured, but the answer, whatever it is,[21] will shed no light on the primary question facing us: What are the *purely* logical requirements governing a device that facilitates the expression of blind truth-endorsements?

1.4 Anaphorically Unrestricted Pronouns

When the question is put this way, a natural "test device" for the role of "true" in English becomes available. We can (artificially) introduce prosentences, or more precisely, impose an additional prosentential capacity on a pronoun already in English ("it") so that it can now also appear in sentential positions but still refer back to quantifiers (in English) that—when functioning as they ordinarily do—only accept anaphora from pronouns in nominal positions (e.g., in our artificial English, we can say, "Some sentence although no one can prove it," or perhaps, "Some sentence, it, although no one can prove it"). Whenever "true" is functioning solely as a disquotational device, it should be replaceable by these *anaphorically unrestricted pronouns*. If a usage of "true" can't be so replaced in a locution, then that use of "true" isn't functioning as a disquotational device compatibly with BTD. I'll call English, absent the truth predicate, but supplemented with anaphorically unrestricted pronouns, *Anaphorish*.

The BTDist, to make good on her deflationism, must show how the use of "true" in explanations and sentences such as (A)–(F) may be replaced by anaphorically unrestricted pronouns in Anaphorish. The point of this exercise is *not* to show that the truth predicate *really is* an anaphorically unrestricted

---

[20] Most react to this sentence as Kirkham (1992, 130) does: "It seems . . . as if the *p* is being used as two different kind of variables within the same formula." This *intuitive* repulsion to the needed generalization is *merely* the manifestation of the bar in English against anaphora crossing syntactic boundaries. Much of David (1994, 61–78) is spent attempting to render formulas similar to this in grammatically acceptable English with the clear implication that a failure to do so counts (somewhat) against their cogency. He is clearly tempted by the van Inwagen (1981) view that an inability to translate a locution into the vernacular counts against its intelligibility. Also see the first epigraph from Quine that opens part I.

[21] Cross-referencing is a well-known sore spot for ordinary language: It's not very good at it—thus the early emergence of schematic letters in mathematics, which otherwise continued in the vernacular. If one wants an explanation for why English, in particular, uses a truth predicate, rather than the forthcoming anaphorically unrestricted pronouns—which cross use/mention divides in the way needed—one need only point to this sore spot.

pronoun (that anaphorically unrestricted pronouns are, in some linguistically respectable sense, part of the "logical forms" of ordinary locutions apparently involving "true" and, say, ordinary quantifiers that range over sentences or propositions), nor is it to show that the meaning of the former devices in some sense amounts to what's expressed by companion sentences wherein the use of "true" has been suitably replaced by an anaphorically unrestricted pronoun. Either possibility is quite remote. Rather, by transliterating sentences, such as (A)–(F), with ordinary uses of the truth predicate (plus quantificational devices) into sentences of Anaphorish, one shows that only the deflationary role of the truth predicate (plus ordinary quantification) is at work in these cases; pending the caveat of the second paragraph of note 11, no more "substantial" understanding of the truth predicate need be involved.

1.5 *Talking Anaphorish*

As we ordinarily speak, when we generalize claims, we say things like:

(4) Everything John said is true,

or,

(5) If John said something, then it is true.

This is nicely captured, in First-Orderese, like so

(6) $(x)(\text{John-said}x \Rightarrow Tx)$,

where both "John-said" and "T" are predicates. Anaphorically unrestricted pronouns—when formalized—allow us to rewrite this like so:

(7) $(x)(\text{John-said}x \Rightarrow x)$,[22]

which can be directly expressed in Anaphorish as:

(8) If John said something, then it.

(8) isn't hard to understand (especially if one puts an appropriate stress on "it"): One makes good sense of it if one recognizes (and accepts) that the "it" in (8) is linked to the quantifier "something," and that nevertheless it's standing (grammatically) in sentential position. I'll also allow cross-referencing of the same sort with names and with variables. Here are two examples:

(9) (10) has four words, and it,

(10) For every sentence A, if John-said$A$ then A.

(9) amounts to the claim that the sentence (10) has four words, and that it's true; (10), on the other hand, has the same content as (4).

One last issue before I move on to truth conditions, proper. The stylistically stolid reader may sulkily presume not to *understand* such things as (8), (9), and

---

[22] See chapter 3, where the formalization in question is introduced.

(10). Such a temper tantrum isn't called for, however: These are the sorts of semi-regimented language one finds scattered throughout, for example, Quine's work.[23] The primary difference between his cases and this one is that the formalism the semi-regimentation is directed toward is different (the forthcoming AU-quantifier formalism, in this case, as opposed to first-order logic); otherwise the strategy is exactly the same. Alternatively, as I've motivated them here, items such as (8) may be seen as *grammatical metaphors* that indicate what the role of the now-missing truth predicate actually was in the original sentence.

1.6 Tarskian-style Truth Conditions

Let's turn, therefore, to the attribution of *truth conditions*. A number of philosophers (and among them, a number of deflationists) are on record as believing that truth-condition theories of meaning are incompatible with a deflationary notion of truth; indeed, many believe that a deflationary notion of truth is incompatible with the giving of truth conditions altogether, except perhaps in an utterly trivial or circular sense.[24] Two reasons for believing the incompatibility thesis are nicely described (but not necessarily endorsed) by Bar-On et al. 2000 (2) as follows:

> First, if ... truth is a flimsy notion, nothing more than a logical device, how can the notion of a condition of *truth* be assigned a significant role in any explanatory theory? Yet truth-condition theories of meaning maintain that the condition under which a sentence is true constitutes (at least part of) its meaning. Second, if ... the truth predicate is just a convenient method of ... semantic ascent, so that speaking of the truth of a sentence, S, is just a way of saying something about the world, then the meaning of "S is true" is parasitic on the meaning of S.... But if so, it would be circular to offer the "truth-condition" of S as part of the explanation of S's meaning.

Keeping the role of "true" in mind (as it was described previously in this chapter) blunts even the prima facie force of these objections. The second objection, as stated, turns on an inference from the role of "true" for semantic ascent (and descent) to the claim that "the meaning of 'S is true' is parasitic on the meaning of S." But this doesn't follow. Even if the meanings of S and "S is true" are linked by principles about quotation and truth, it still needn't follow (and actually, it's hard to see why it *should* follow) that the meaning of "S is true" is "parasitic" on the meaning of S.[25] Apart from this, the second objection seems ominously similar to that charge of circularity Tarski (1944, 356–57) rebutted long ago. In brief, it's that there is a purported problem

---

[23] E.g., Quine 1960, chapters 4 and 5.

[24] For a list of deflationists who believe one or another version of this incompatibility thesis, see Patterson forthcoming(a), note 1.

[25] Consider the meaning of "Pa," and the meaning of "$(\exists x)(Px \ \& \ x = a)$." Only on certain (controversial) theories are these the same, despite the interdeducibility of the sentences in question.

with the truth conditions he gives *for* the connectives, since such connectives themselves appear *in* the truth conditions given. In general, this *is* problematic if the truth conditions are supposed to be, say, things that (for some reason or other) *aren't supposed to use* the notions that the truth conditions are for. All by itself—without, that is, additional constraints on what a theory of truth conditions is supposed to do—there can't be an objection to using notions *in* a characterization of the truth conditions *of* those notions. Such characterizations, of course, *could* prove utterly trivial—but triviality isn't guaranteed, and whether triviality results can be decided only by an inspection of the characterizations themselves. This relates to the first objection: Given that "true" is a piece of the logical apparatus, there can hardly be a complaint about its use in the characterization of truth conditions, any more than one should complain, say, about the use of the connectives or quantifiers themselves in such characterizations. One *might* complain about the central use of the word "truth" in the nomenclature "truth conditions." But from the BTD point of view, even this terminology can be argued to be innocuous: "True" points through to the sentences (or propositions) themselves: Truth conditions, in turn, are conditions under which certain sentences are true or false given the truth and falsity of other sentences, or, as neatly expressed in Anaphorish, they are conditions under which *certain sentences*, given *other sentences*. And it's entirely reasonable to regard the interlocking of sentences, in the way that truth conditions interlock them (in, for example, the standard Tarskian approaches), as part of the meaning of such sentences.

I've followed the version of the concern given by Bar-On et al. 2000, which focuses on the functional role of truth in semantic ascent and descent, and which worries about the compatibility of that function with truth-condition theories of meaning. The suggestion I'm ultimately making is that if we look at typical truth-condition clauses, we'll see that they transliterate easily into Anaphorish. If that's correct, then the success of the transliteration shows that only the deflationary use of "true" is at use in such clauses that everyone who takes truth condition theories of meaning seriously already regards as successful.

Before exhibiting such transliterations, however, it's worth noting other versions of the circularity concern that don't focus on the functional role of truth in semantic ascent and descent. Dummett (1959, 7), for example, using the jargon of the redundancy theory of truth, writes:

> The conception pervades the thought of Frege that the general form of explanation of the sense of a statement consists in laying down the conditions under which it is true and those under which it is false.... But in order that someone should gain from the explanation that *P* is true in such-and-such circumstances an understanding of the sense of *P*, he must already know what it means to say of *P* that it is true. If when he enquires into this he is told that the only explanation is that to say that *P* is true is the same as to assert *P*, it will follow that in order to understand what is meant by saying that *P* is true, he must already know the sense of asserting *P*, which was precisely what was supposed to be being explained to him.

Dummett (naturally) begins the next paragraph with: "We thus have either to supplement the redundancy theory or to give up many of our preconceptions about truth and falsity." Identification of the meaning of the right and left wings of the T-biconditionals, or treating the T-biconditionals as providing the "meaning" of the word "true" really can land us in a circle should we take truth-condition theories of meaning seriously. But no circle is evident merely on the grounds that the function of the truth predicate is semantic ascent and descent.

Horwich (1998, 68) puts his version of the objection this way:

> Understanding a sentence . . . is a matter of appreciating what must be the case for the sentence to be true—knowing its *truth condition*. That is to say, one must be aware that "Tachyons can travel back in time" is true iff tachyons can travel back in time. Therefore it is not possible to agree with the minimalist claim that this knowledge also helps to constitute our grasp of "is true". For in that case we would be faced with something like a single equation and two unknowns.

As I read this objection, it is that "'Tachyons can travel back in time' is true" is playing two roles (and it can't). The first is that it constitutes (part of) our knowledge and understanding of the truth predicate, and the second is that it gives the truth conditions for the sentence "Tachyons can travel back in time." Here, clearly, it isn't the semantic ascent and descent role of the truth predicate that's causing problems—that role is actually a way out of the problem being posed. For the problem is due to the claim that our understanding of that predicate is constituted by the T-biconditionals, and those who take their understanding of the truth predicate to issue from its semantic ascent and descent role—rather than from the T-biconditionals—can conveniently reject this claim.

Two concerns remain. The first is the triviality worry raised above—that truth-condition clauses, containing the same logical terms in both wings of the biconditionals, will prove uninformative about meaning. The second is the question about whether the role of "true" in such clauses is solely as an enabler for blind truth-endorsements. Consider the following paradigmatic truth conditions, one repeated from above:

(G) For all sentences *A* and *B*, (*A* v *B*) is true if and only if either A is true or B is true.

(B) For all sentences *A* and *B*, (*A* & *B*) is true if and only if *A* is true and *B* is true.

Both of these are easily recast in Anaphorish without the truth predicate:

(G') For all sentences *A* and *B*, (*A* v *B*) if and only if either *A* or *B*.

(B') For all sentences *A* and *B*, (*A* & *B*) if and only if *A* and *B*.

Tarski's truth clauses (of which (G) and (B) are examples) are, as I've said, paradigmatic examples of truth-condition attributions for a class of sentences of certain forms (ones with the primary connective being the disjunction in the case of (G) and the ampersand in the case of (B)). What are truth conditions, and these truth conditions in particular, supposed to do for us? As mentioned earlier, a popular construal is that the truth conditions of a sentence are supposed to tell us the circumstances under which that sentence is true or false. This way of putting the matter will certainly strike some as uninformative because it's compatible with the unadorned list of T-biconditional for *every* sentence: "*S*" *is true iff S*. What's often added, as noted, is that such items are valuable, but only in the context of a semantic theory that gives us a great deal more, gives us in fact what's often described as "compositional semantics." Again, Tarski provides the paradigm. His truth clauses (of which (G) and (B) are examples) explain how the truth (and falsity) of sentences are due to the truth and falsity of their subcomponents (broadly construed)[26] of sentences. If this is all that's needed, notice that the resources of Anaphorish clearly suffice, as (G') and (B') make clear. Speaking of the truth and falsity of sentences in terms of the truth and falsity of their subcomponents translates easily into speaking of those sentences in terms of their subcomponents, as (B') has it, or, in Anaphorish without schematic letters:

(H) If two sentences are conjoined by an ampersand, then *that* conjunction iff the first sentence and the second sentence.

The point is easy to see, even if English must be tortured into Anaphorish to manage it. Speaking of the connection of the truth or falsity of sentences to the truth or falsity of other sentences requires a recursive characterization of the truth and falsity of all sentences via an ordinary quantifier that ranges over sentences; but such a recursive characterization is available in terms of quantification over sentences and a device—an anaphorically unrestricted pronoun—that can stand stead in both nominal and sentential positions.

An inspection of the truth conditions that the standard Tarskian approach offers for the logical apparatus, the logical connectives and the quantifiers, shows those truth conditions to be nontrivial—despite the appearance of the same logical items on both sides of the clauses—because it allows a recursive characterization of sentences in terms of semantically significant subcomponents of those sentences.[27]

A version of the triviality worry remains despite the above analysis. For, recalling Horwich's " 'Tachyons travel back in time' is true iff tachyons travel

---

[26] "Broadly construed," because Tarski's approach, and others modeled on it, supply the truth conditions of sentences with quantifiers in terms of the *satisfaction* of subcomponents that needn't themselves be sentences. Details about how this goes doesn't affect the discussion here, and so I'm leaving it aside. I'm also leaving aside—but only until chapter 4—issues having to do with the stratification of languages that the Tarskian approach brings with it.

[27] The clauses may still seem trivial if one focuses only upon the appearance of the clauses instead of noticing how these clauses interlock *all* the sentences of the language so characterized— something that becomes transparent as soon as it's realized that (G) and (B), and their kin, describe a *class* of sentences of arbitrary complexity. This, however, is the point of the phrase "recursive," in Tarski's characterization of what his truth conditions offer.

back in time," we can notice that this gives the truth conditions of "Tachyons travel back in time" not by that sentence interlocking with other sentences, but directly. Is this truth condition, therefore, a trivial one? Well, no. An English speaker knows it by virtue of grasping the meaning of "true" and quotation, but nevertheless, the two wings of the biconditional aren't analytic equivalents as a result of that fact. Is this truth condition—where a quoted sentence, and that sentence unquoted appear in each wing of the biconditional—genuinely *informative*? It's easy to see why people have thought otherwise.

1.7 The Base Clauses of Tarskian-style Truth Conditions

"Tachyons travel back in time," is true iff tachyons travel back in time is an example of the sort of base clause needed in a Tarskian-style truth-condition theory of meaning when sentences don't have logical constants in them. As we've seen, these typically take the form of (A), here repeated:

(A) "Snow is white," is true iff snow is white.

It's not clear how this should be transliterated into Anaphorish. One possibility is this:

(A') There is something that is "snow is white" and it iff snow is white.

This is obviously true for English speakers able to get the hang of Anaphorish. Although (A) and its ilk have been accused of being inadequate for the purposes of supplying truth conditions, they *are* adequate, of course, for the purpose of fixing the truth predicate so that it facilitates semantic ascent and descent;[28] but more than clauses like this—it may be thought on the basis of the closing remarks of 1.6—are needed for describing the conditions under which such sentences are true and false.[29] What more is this? Well, one needs to give the conditions that explain under what circumstances "snow is white," *is* true. Roughly speaking (on one view), one needs to point out that the item that "snow" refers to has the property "white" refers to; maybe even more than this is needed: Perhaps one needs to explain how it is that "snow" refers to snow (and "white" to white); in short, perhaps one needs to expose the *mechanisms* of reference. And, in turn, maybe this requires an analysis of the sort of causation that's involved in how reference is passed among

[28] At least they are when blind truth-endorsements are restricted to one's own language or idiolect, and provided one tells a successful story about how T-biconditionals handle ambiguity and sentences with demonstratives. (See chapter 2, where some of these issues are raised.)

[29] Field (2001c) essentially raises this issue—although in the guise of a concern over the incompleteness of Tarski's theory of truth. I should add that, strictly speaking, the above has shifted to a debate over whether a truth-condition theory of meaning suffices as a theory of meaning, and so this goes beyond the question of whether the BTDist is right about truth condition theories of meaning; for it's already been shown that anaphorically unrestricted pronouns can do the work that "true" does in such theories. Despite this, the concern is worth pursuing now because it can be shown that even richer views about what's required for a theory of meaning are entirely compatible with uses of "true" restricted to its semantic ascent and descent functions.

members of a community of speakers; that is, perhaps social facts about how reference is fixed must be brought into play. In any case, so this opponent of the sufficiency of such trivial base clauses may say, what we want from "truth conditions"—if they are to play the role in meaning and content attributions that we need them for—is entirely eviscerated if the deflationist has only the T-biconditionals to offer by way of elucidation of the truth conditions of "snow is white," "grass is green," and so on. And what *this* is supposed to show, at the end of the day, is that truth conditions, *if done right*, go beyond the deflationist construal of "true."

Recall that the careful BTDist requires that every use of "true"—even when giving truth conditions—involves it only as a device of semantic descent. Let **R** be a substantial characterization of the truth conditions of sentences (without logical vocabulary within them) along the lines just described, and let **B**x be a one-place predicate holding of sentences of a certain form (say, n-place predicates concatenated with n terms). Then consider the following truth-condition offering:

(11) $(x)(\mathbf{B}x \Rightarrow (\mathbf{R}x \Leftrightarrow Tx))$.

Can this be recast anaphorically? Of course:

(12) $(x)(\mathbf{B}x \Rightarrow (\mathbf{R}x \Leftrightarrow x))$,

where now the quantifier "(x)" quantifies into both nominal and sentential positions because the other instances of "x" in the formula are understood as formalizations of the anaphorically unrestricted pronoun "it." Or, putting the claim directly into Anaphorish: For all sentences constructed of an n-place predicate followed by n names: **R** holds of it iff it.

That is, nothing stops the BTDist from giving truth conditions as substantial as you please for sentences: Doing so is *totally* compatible with a truth predicate whose only role is semantic ascent and descent. (That's what rewriting (11) as (12) indicates.) No doubt proponents of the incompatibility thesis may feel cheated by the foregoing. But why? Perhaps because all along, the issue for them has *not* been the role of the truth predicate; that's been but a distracting sideshow. The real issue is the vaguer one of *whether truth has an underlying nature or not*; that is, such deflationists aren't (*really*) concerned with BTD but with MTD!

Actually, before turning to that (thorny) issue, I should confess that something else has happened that may make (certain) deflationists feel cheated. The *biconditional* truth deflationist may feel that all that's allowed is a notion of truth governed by T-biconditionals, and such a deflationist may protest that, **R** as understood, is hardly restricted to *those*. But so what? A main proponent of this sort of view, Horwich (1997, 95–96), writes:

> The basic thesis of deflationism . . . is that the disquotation schema . . . is *conceptually fundamental*. . . . [O]ur overall deployment of the truth predicate—the sum of everything we do with the word "true"—is best explained by taking the *basic* fact about its use to be our inclination to accept the instances of the disquotation schema.

I don't think the BTDist need accept this, as I argued in 1.1, but—leaving my objections aside—where in the foregoing has even this stricture been violated? The point of the disquotation schema is to allow semantic ascent and descent; and Horwich (1997, 97) concedes this. And so it seems there can be no objection to the giving of truth conditions **R**—substantial truth conditions—to sentences. Of course, if we'd found ordinary uses of "true" that clearly required an interpretation in terms of **R** rather than in terms of its semantic ascent and descent roles, then the BTDist would be in trouble: Such uses would be a counterexample to her thesis. But what's been found instead (if the MTDist is wrong, let's say) is that, compatibly with the role of "true" as a device of semantic ascent and descent, the sentences that "true" as a predicate holds of are ones that in fact **R** holds of as well. But this doesn't bear on the use of "true" in the language, including its use in giving (11), for it operates there only in its semantic ascent and descent role. Furthermore (at least so far) in discovering its coextensiveness with **R**, no attempt has been made to undercut the so-called foundational status of the T-biconditionals. In fact, it's confused to think that the giving of substantial truth conditions ever *could* undercut their status.[30]

1.8 Theories of "True" and Theories of Truth

Why? Well, it helps to make a distinction between a theory of "true" and a theory of *truth*. A theory of "true" is a theory about a piece of language ("true") and its (indispensable) role. A theory of *truth* is a theory, if such is possible, about the systematic uniformities (if any) among *truths*. These two sorts of theories are, we've discovered, sensibly *separated* from each other. BTD is a theory about "true"; that's why it's compatible with any number of theories about *truths*—only *one* of which is MTD.[31]

So what about truth, and a theory of *it*? We've seen indications of what a deflationist theory of "true" is supposed to look like; what's a theory—any theory—of *truth* supposed to look like? A theory of truth, as I understand it, isn't a *list* of truths, or even a recipe for generating such a list; rather, it's a theory of how, if at all, truths are structured. One might try to characterize theories of truth in terms of views about the *property* of truth: E.g., substantial theories of truth hope to explain the property of truth in terms of something else; and it's also a theory of truth, as I understand it, to deny that there is any such property of truth to explain.[32]

---

[30] What *will* undercut their status, as will be shown in chapter 2, is the ordinary application of "true" to sentences in other languages. But for the moment I'm allowing the assumption that the truth predicate is fully enabled in its semantic ascent and descent roles by means of T-biconditionals.

[31] I sense sympathy with this distinction, or something like it, in Grover 2002 and Devitt 2002. I sense hostility to the distinction in David 1994, Lynch 2000, and Sher 2004.

[32] I owe the suggestion of characterizing theories of truth in terms of explaining the *property* of truth to Douglas Patterson; but I have to add that I'm somewhat uneasy about this construal of such theories. Some might take a dim view of properties, and yet have a substantial correspondence theory of truth nevertheless.

A *substantial* theory of truth *could* look very like a (deflationary) Tarskian theory of "true." Here's how. Start by adopting this metaphysical view: There are objects and there are properties, and the former have (some of) the latter. Furthermore all such objects fit neatly in a domain.[33] Then one might think that all truths arise recursively, Tarski-style, given such a domain. That *really is* a theory of *truth*; indeed, it has a right to be called a "correspondence theory of truth," if not "*the* correspondence theory of truth": Statements without logical terms are true if they correspond to the instantiation facts about relations (and objects instantiating such relations). Statements with logical terms are, of course, true or false based on the Tarskian truth-clauses.[34]

I'm going to call this *the correspondence construal* of Tarskian semantics; and although I won't try to do this, I think that a textual case can be made that Tarski had something like this interpretation of his approach in mind when he wrote his 1932 (1983a), but that he deserted it for the neutrality interpretation in his 1944. My point in offering it now, however, is just to give an indication of what a (substantial) theory of truth might look like. The reader familiar with the various theories of truth—correspondence, coherence, and so on—no doubt realizes that I think these are *at best* theories of truth rather than theories of "true"—"at best" because in some cases they aren't even that: They're epistemological theories *disguised as* theories of truth. Similarly, much of the work on the semantics of names and kind terms (associated with Kripke, Putnam, and others) belongs properly to (a substantial) theory of truth rather than to a theory of "true."[35]

I should at least give an indication for why I think nothing in general can be said about truths—why there really isn't (really can't be)—anything like a (substantial) theory of truth. To do so, I must bring up considerations argued for elsewhere (see Azzouni 2004a), and be very sketchy about them here. I presuppose a metaphysical claim: Nominalism is correct (there are no abstracta). Nevertheless, mathematical statements are *true* and are intertwined with ordinary empirical statement (about things that *do* exist) in such a way that no semantic theory is possible that separates statements that are true (and are solely about things that exist) from statements that are true (and are—at least

---

[33] This is a nontrivial assumption because there might be *too many* objects to fit comfortably in a domain. See Field 1989, 31–32, for the set-theoretical version of the worry. There are, of course, ways around this for the metaphysician: One is to claim that the actual world is modeled by a(t least one) model that possesses the same (appropriate) logical properties that it does. Since the concern in this chapter is oblique to a defense of the metaphysician, I leave further discussion of this issue aside.

[34] I'm helping myself liberally to the truth idiom since I'm writing in English; but it should be clear that all of this can be said in Anaphorish; nothing being said involves a use of "true" that goes beyond its semantic ascent and descent role.

[35] I'm papering over an issue here: Which parts of the theory of truth belong to semantics proper, and which don't—which belong instead to pure metaphysics, say? (And which parts belong to *both?*) These are, in general, fairly subtle matters. This much can be said now: Just because the theory of truth isn't about the semantics of "true" doesn't mean it doesn't contain *semantics*. The latter issue turns on the question of how the subject of semantics should be demarcated. And, as I said, I don't think this is an easy question *at all*.

partially—about fictions). Consequently no general theory of truth—at least of a correspondence sort—is forthcoming. This hardly shows, of course, that no *other* theory of truth is forthcoming—even granting my nominalism; for, no doubt, readers can think of (or anyway, recollect) other ways that truths can be seen as having uniform properties. I'm sceptical, however, that these other options will work because I'm sceptical that, once we desert the correspondence option, there are any genuine uniformities to attribute to the *truths themselves*, as opposed to what may be broadly described as the epistemology of truth: how we establish truths, or supply evidence for them, or something like that. In any case, I can't say any more about this now.[36]

Here, therefore, are some summary remarks about BTD, MTD, and the vexed issue of truth conditions. Truth conditions are unproblematic for the BTDist in any case. Nontrivial truth conditions are also available for the MTDist, at least when it comes to truth conditions that interlock statements of our language with other statements; this is illustrated by the Tarski truth conditions for the connectives and the quantifiers. The MTDist, however, is likely to claim that nothing more than simple truth conditions are available for sentences without logical terms such as, "Grass is green," "Sally is between Jack and Jill," namely: "Grass is green," is true iff grass is green, and so on. His substantialist opponent, if he doesn't think these suffice, can try to offer more substantial truth conditions along the lines of **R** for such statements. As a result the MTDist and the BTDist may part ways, for the latter can accommodate such a substantialist view about truth, although the former won't abide it. The MTDist will deny there is anything to say (in general) about how such statements are made true or false. In some cases, he will say, such statements are made true or false by truthmakers and their machinations (electrons and their properties, say)—and for these specific cases he might be quite sympathetic with causal theories of reference, say, or such theories modified in certain ways;[37] but in other cases—e.g., number-theoretic truths—he is capable of denying the existence of any such truthmakers; and he can claim that the existential statements about numbers are made true in a way that doesn't involve correspondence in any sense, however lightweight. The MTDist won't deny that there are various stories to tell about how these different statements are made true and false, and he may be serious about metaphysical commitments in certain cases as a result; but he won't think this has much to do with a (general) theory of truth, or with a property in some substantial sense that all true statements share. Perhaps this is ultimately

---

[36] But might it be suggested that the presumption of a uniform theory of truth (an underlying metaphysical nature for all sentences or propositions that are true)—something in fact like Tarski's theory of truth construed as a substantial correspondence theory—coupled with compelling evidence that some sentences about sheer fictions are nevertheless true, is what drives some philosophers (e.g., T. Parsons (1980), Routley (1980), Deutsch (2000)) to what can only be described as detailed metaphysical studies of nothing? That is, to metaphysical arguments that such and such properties—and not others—hold of what are (also) described as utterly nonexistent entities?

[37] See Azzouni 2000b, parts III and IV, for one way this could go.

a matter of nomenclature. One might argue that a "theory of truth" can say very different things about different classes of truths, and so strictly speaking, describing the MTDist view the way I do is still compatible with taking it to be a substantial theory of truth in that sense.[38] At the moment, I'm only stressing the disagreement with that philosopher who wants to tell a uniform story about *all* truths—the correspondence theorist, for example—because views like that have been so prominent in the literature, both in their own right, and as foils for deflationism broadly construed.

My partiality to MTD shouldn't cause the reader to underestimate the independence of BTD from doctrines about *truth*. I've stressed the compatibility of BTD with MTD—the doctrine that there is nothing in general to say about truths—where the foil to MTD is some sort of global correspondence theory.[39] But BTD, in fact, is compatible with a far more drastic view: *that there are no truths at all.* Imagine that what we *take* as true is determined by *and only by* coherentist factors involved in how we make our empirical observations fit in with our background beliefs; and suppose these factors won't cause a convergence upon a unique, or even particularly stable, set of truths. Imagine, that is, that science (really) is an endless cycle of theorizing in which we're sure of a set of truths for a time; but over long periods of time, as we carry out additional empirical studies, no truth ever stays in the (purported) truth-set for good. (Call this "the fallibilist's nightmare.") And suppose, even worse, that this isn't a mere matter of epistemology; rather, metaphysically speaking, there isn't anything "out there" that favors one set of truths over another: That is, there are (metaphysically speaking) no truths at all. In such a case, the appearance of a stable set of truths—at a time—would be due only to the parochial grip we had on the world at that time. Regardless, we would still need blind endorsements, and so we would need a truth predicate (or its equivalent); but it wouldn't be that the truth predicate (really) picked anything out. It would appear to have a (settled) extension (at a time)—whatever extension, that is, that at a time we took it to have; but in fact it would have, at best, ever-changing and shifting extensions-at-a-time. This state of affairs, as

---

[38] Lynch 2000 seems to have a view like this; but he also seems to think his view involves a theory of "true" or at least "that it is difficult to distinguish an investigation of the (deep metaphysics of the) property of truth from an investigation of the concept" (212). If I understand this claim correctly, it denies a neat distinction between a theory of truth and a theory of "true."

I should say, though, that a view that separates truths into separate categories according to different theories of truth that apply to them presumes, to begin with, that truths *can* be neatly sorted into various categories, and I've already given reasons to doubt this is possible. In part II of Azzouni 2004a, I illustrate in some detail how pure mathematics interpenetrates with the empirical (in certain applications of physics) in a way that makes it impossible to sort the resulting doctrine into "purely" empirical truths and "purely" mathematical ones. This interpenetration phenomena among truths is quite widespread—indeed, it's a form of holism.

[39] A view, that is, that either takes all (true) sentences (or propositions) or well-defined classes of such (empirical language excluding fictional and mathematical discourse, for example) to correspond to structured entities (facts, objects, relations) of some sort. MTD is compatible with a local sentence-by-sentence view of correspondence: that each (true) sentence must be judged on a case-by-case basis to determine what, if anything, it corresponds to.

I said, is totally compatible with the BTDist's view of "true"; but (of course, as I've said) a metaphysically rich correspondence view is equally compatible with BTD.[40]

I should also stress and amplify the remarks of note 11 above, because they bear both on the metaphysical neutrality stance of the BTDist and on the possibility of one's being both a BTDist and an MTDist. This is that whatever story the BTDist ultimately tells about how devices for blind truth-endorsement work, it can't presuppose notions that would violate the metaphysical neutrality that the BTDist is committed to. I thus understand the BTDist neutrality stance as an important part of her "deflationism." So what's needed for the ultimate success of "deflationary exegesis" is not *just* that "true" be only a device of semantic ascent and descent, but that the theory of how that device works itself doesn't require a commitment to *more* than that of truth as a device of semantic ascent and descent. It's worth noting that some—in light of what's required to actually get a device to do what's needed for semantic ascent and descent (see chapters 3 and 4)—may think such a substantive commitment is worth reconsidering. (I'll say something more about this in the conclusion to part I.)

1.9 Concluding Remarks

I've distinguished between two sorts of deflationism: biconditional truth deflationism (BTD), which is a doctrine about the truth predicate in the vernacular that claims the use of that predicate is restricted to semantic ascent and descent, and metaphysical truth deflationism (MTD), which is a doctrine about truth that holds there is nothing of interest that all truths have in common. I've separated BTD from other assumptions often chained to it, doctrines about the centrality of T-biconditionals, either to our conceptual understanding of truth, or as trivial analytic entailments recognizable because their left and right wings mean the same. However, despite the cleaning up of the BTDist position that I've engaged in, I'll nevertheless defect from the position in chapter 2, for we will discover that there are uses of the truth predicate that—although still connected to its function as a device for navigating between use and mention—can't be captured by means of T-biconditionals. Instead, the anaphorically unrestricted pronoun—used in this chapter merely as a symptom of the deflationary role of the truth predicate it replaces in translations of sentences from English to Anaphorish—will become the backbone of a distinctive approach to truth that will enable talk of truth to transcend restrictions to our own language that the BTDist must respect.

---

[40] Some might resist this thought experiment by claiming that the T-biconditionals *require* an unchanging extension for the truth predicate. It's hard to see what argument could yield this conclusion without simultaneously faulting our ability to use the truth predicate to assert claims about statements that we can discover ourselves to subsequently be wrong about. I don't deny, of course, that we (or some of us) have strong intuitions about the stability of truths—but that's different from trying to read a stability condition on truths off of the T-biconditionals themselves.

I also, by means of the distinction between BTD and MTD, have tried to pry apart philosophical concerns about the nature of the truths that fall under the predicate "true" as opposed to the role of the predicate itself. There are reasons to embrace MTD, as I've indicated, but these reasons lie apart from the question of the role of "true" in the vernacular.

Finally, I've done *some* of the work necessary to showing that, indeed, the truth predicate does only have a semantic ascent and descent role in the vernacular, by showing—contrary to the impression of a number of philosophers—that nothing about that restricted role prevents the truth predicate from functioning as needed in truth-condition theories of meaning.

One last point about nomenclature. I've endeavored to distinguish between theories of truth and theories of "true." Unfortunately, this distinction isn't widely drawn, and discussions in the literature invariably make reference to "theories of truth" in which the distinction noted is implicitly, if at all, made—that is, sometimes a theory of "true" is meant, and sometimes a theory of truth is meant. Worse, in some cases, such as "Tarski's theory of truth," one faces a great deal of work trying to determine whether what's of concern is his theory of "true" or his theory of truth. Because of the previous point—and because some philosophers (recall note 31) are likely to reject the distinction altogether—in what follows I won't use a nomenclature that distinguishes theories of truth from theories of "true." That is, I'll often follow the literature (e.g., in chapter 2) in concerning myself with "Tarski's theory of truth" so called without pausing, say, to rewrite "truth" as " 'true.' " Where the distinction clearly bears, of course, I'll revert to it. But almost everything in the remaining chapters of part I is concerned with a theory of "true," not a theory of truth, and so there is unlikely to be a problem if, for example, the reader who likes the distinction drawn in 1.8 understands the discussion in chapter 2 of "Tarski's theory of truth," as a discussion of "Tarski's theory of "true."

2

The Transcendence
of Truth

2.1 The Immanence of Tarski-Truth

Quine (1970, 19) draws a distinction between immanent linguistic notions and transcendent ones: "A notion is immanent when defined for a particular language; transcendent when directed to languages generally." Truth, explicated Tarski-style, proves an immanent notion in just this sense. The reasons seem obvious: Tarski's approach starts with an interpreted language **L** ("object language") already in place that we'd like to provide a truth predicate for. An adequacy requirement—Convention T—that Tarski (1983a, 155) places on any truth theory (described, traditionally, as occurring in a "metalanguage"[1]) is that for each sentence of **L**, the statement *x is a true sentence if and only if p* is derivable from it. This schema is understood as one in which any sentence of **L** is substituted for "*p*" while an individual name of this sentence replaces "*x*." Tarski-style approaches define or axiomatize truth by (i) giving syntactic structural-descriptive names of the sentences **L** (via a lexicon, recursive rules of construction of well-formed formulas, and so on), (ii) recursively characterizing satisfaction of well-formed formulas f in terms of subformulas of f, and (iii) taking truth as the restriction of satisfaction to closed formulas.[2] By means of

---

[1] Tarski's construction of a truth predicate for a language, under certain circumstances, needn't occur in a different language.

[2] Notice that the axiomatization or definition of Tarski-style truth thus piggybacks on a syntactic characterization of the sentences of **L**. There are technical reasons, connected to the resources of the metalanguage and those of the object language, why sometimes a definition of

structural-descriptive names for each sentence l of L, l is associated with a sentence l^* of the metalanguage that can be shown to be a suitable semantic proxy for it. Indeed, precisely that semantic suitability is the content of the biconditional theorems that follow from the resulting theory of truth: The right side of the biconditional is l^*, and the left side is the statement that the sentence l (characterized by its structural-descriptive name) is true.

The approach yields an immanent notion of truth in two senses. First, because the resulting truth predicate only applies to sentences of L, but second, and more significant, because the resulting notion of truth is beholden to the specific syntactic properties of the language L that it's defined or axiomatized in terms of. As a result, Tarski's full "theory of truth" contains much more than what's needed to fix the truth values of blind truth-endorsements: It contains, as mentioned, a theory of the syntax of the language L it's a truth theory of, and, in addition, it contains the compositional truth conditions for L (sentences of the form (B) and (G) from chapter 1); it also contains resources for defining the proof theory of a theory in the language of L. All of these elements participate intimately in the definition, or axiomatization, of truth— Tarski-style—and so Tarski's definitions or axiomatizations must be modified on a case-by-case basis for languages containing new idioms or syntactic constructions not explicitly considered by him.[3]

There is no doubt that all this is *invaluable*; but is there any reason, other than that Tarski's groundbreaking paper provided *all* of it in one package, to regard these latter items as crucial to a theory of *truth*? A number of philosophers have presumed that the additional generalizations available via Tarski's methods *are* a necessary component of a theory of truth:[4] Ketland (1999, 85) mentions several typical metalogical truths and explicitly requires an "adequate" theory of truth to contain them: "For any closed formula ϕ, $\neg\phi$ is true if and only if ϕ is not true; If Σ is a set of true closed formulas, then any deductive consequence of Σ is true; For any set of closed formulas Σ, if Σ is true, then Σ is consistent."[5]

Such truths go beyond what is proper to a theory of *truth*. If such a theory is supposed to fix the truth values of blind truth-endorsements, then such claims certainly go beyond doing *this*. Of course such claims, as well as being semantic truths, are *also* blind truth-endorsements; but *that* (all by itself) hardly implies they should be included *in* the theory of truth.

Tarski-style truth is available in a metalanguage for an object language, and why sometimes only an axiomatization is available. This isn't germane to the concerns of this chapter, and so I'm omitting further discussion of it.

[3] This can't always be done in a straightforward way. An example of how *ingenious* extensions of the Tarski construction sometimes have to be to handle new logical idioms is that of modal terminology, *necessity* and *possibility*, and the sorts of semantics that's been designed for them.

[4] E.g., McGee (1993), Gupta (1999), David (1994), Heck (1997), Shapiro (1998), and Ketland (1999).

[5] Ketland (1999, 90) also writes, "Any *adequate* theory of truth should be able to prove the 'equivalence' of a (possibly infinitely axiomatized) *theory* T and its 'truth' $True(T)$ (that is, the metalanguage formula $\forall x(Prov(x) \rightarrow Tr(x))$)." For similar sentiments, see Shapiro (1998).

It may seem that not including such semantic principles in a theory of "truth"—on the grounds that they go beyond the strict letter of what's needed to fix the truth values of blind truth-endorsements—is mere nomenclatural prissiness. After all, semantic principles (as just noted) *are* blind truth-endorsements, and if it's the job of any theory of "truth" to fix the truth values of blind truth-endorsements, then it will fix the truth values of semantic principles as well—and so any theory of "truth" *should imply* such truths—just as Tarski's full theory does, and just as Ketland and others claim that any "adequate" theory should.

This last bit of reasoning overlooks the fact that there are *other* languages where the syntactic and semantic principles are different (e.g., languages governed by alternative logics), but where blind truth-endorsement is *still* needed. In these cases, semantic principles, such as those about connectives— e.g., "&"—may well be different or missing, and yet *the blind truth-endorsement device is the same one.* Such a variety of existing languages should incline us to try to logically segregate what's needed for blind truth-endorsement from other particular blind truth-endorsements—be they generalizations about the semantic properties of languages or generalizations about other aspects of sentences. Having said this, I should add that whatever way is chosen to characterize the blind truth-endorsement device, it should be one that naturally admits augmentation with semantic principles (of one sort or another).[6]

2.2 The Transcendence of Vernacular Truth

Perhaps for reasons peculiar to Quine's views about proper names, and even though he was the first to stress the value of the truth predicate because of the generalizations about inference and semantics that it enables us to capture, Quine (1970) nevertheless underdescribes the role of the truth predicate in the vernacular; and this tendency inertially continues among contemporary philosophers sympathetic to the disquotational line.[7] For Quine's focus on generalization has the effect of (i) inducing a general failure to realize just how widespread use of ineliminable blind truth-endorsements in the vernacular is— especially endorsements involving singular terms that can only dubiously be treated as generalizations, e.g., "Maxwell's theory of gases is true," "Fermat's last theorem is true," "Newton's theory of the moon is false," and so on—and because of this focus on generalization, (ii) naturally turning us toward the consideration of technical tools for capturing the role of truth, such as sentential substitutional quantification or infinite conjunctions (recall 1.3), that—as I'll

---

[6] Gupta (1999, 363), apart from also arguing that a theory of truth should imply facts about logical apparatus, offers other simple facts he thinks a theory of truth should explain: (i) The Moon is not true; and (ii) There are as many truths as there are untruths. Again, I see these as substantial claims—even if obvious ones—and I don't see that a theory of truth is required to imply them; although they can easily be part of a theory that includes a theory of truth along with other substantial (general) claims *about* truths (and their vehicles).

[7] See, for example, Leeds 1978; Williams 1988, 424; Horwich 1999, 241.

show in the course of this chapter (2.8–2.10)—are no more successful in capturing how blind truth-endorsement operates in the ordinary vernacular than Tarski's original approach is.

In this section, I review the role of the truth predicate in the vernacular,[8] and show how its transcendence naturally follows from that role. To begin with, notice that blind truth-endorsements that we can't (pragmatically) do without, and that are at best only implicitly general, arise in the vernacular regularly. "Peano's axioms are true," "The first sentence the current president of the United States uttered upon taking office is false," "Einstein's theory is true," "Newton's theory of gravitation is false." These (or most of them) are blind truth-endorsements that responsible epistemic agents can reasonably assert even when those agents can't replace the endorsements with explicit statements. Indeed, it's important to stress how *crucial* it is that we can blindly endorse statements available to us only by name or description. For example, members of Congress (and other political administrators) often have to argue for programs on the basis of the truth of scientific doctrines that they neither understand nor can even *state* coherently.

But such blind truth-endorsements—by speakers ignorant of the actual statements endorsed—are hardly restricted to politicians and the like. Because of the division of knowledge that our linguistic community currently enjoys, it's routine for quite sophisticated scientists to rely on implications to their discipline of theories from other disciplines that they are otherwise unfamiliar with. The implications they actually use may thus be items they *can* easily state although this needn't be true of theories themselves that the results follow from.

Field's (2001a, 122 n) deflationist argues that we don't make blind truth-endorsements to statements we don't "understand"—more specifically, "we don't now understand attributions of disquotational truth to sentences containing future words." The claim is implausible—as a description of vernacular truth-endorsement—if only because it places as a condition on the understanding of a blind truth-endorsement that one understand what's being endorsed: And this is at odds with our practice. *A* has a famous mathematician *B* staying at his home. One day *B* says to *A*, "I've shown Goldbach's conjecture." *A* knows no mathematics (although he knows that his childhood friend *B* is now a famous mathematician, that B is sober, honest, never bluffs, and he also knows what it means for an important mathematical result to be established—knowing *all this* is, of course, entirely compatible with knowing no mathematics). Later, while *A* is sitting with other mathematicians (waiting for *B*, say, at a restaurant), Goldbach's conjecture is mentioned. "It's true," *A* says. "You don't even know what it means," someone rudely retorts. "Maybe not," replies *A*, "but *B* is staying at my home, he told me it's true, and I have every reason to believe him." (And now so do *they*.)[9]

---

[8] I previously did so in chapter 1 of Azzouni 2004a.

[9] Note the point: If Field's (2001a) deflationist's claim about blind truth-endorsement accurately described our practices in the vernacular, then the natural response to *A* would be one of bewilderment—a request to make clear what he'd meant by his peculiar use of "true"—or a

Given the foregoing, it should be no surprise that part of our truth-endorsement practices is not only the right to endorse or deny statements we don't understand and can't express directly (if we know on other grounds that they are true or false), but also the right—when similarly justified—to blindly endorse or deny statements uttered in *other languages*. We're often, that is, in a good (epistemic) position to endorse or deny a statement that we can't understand because we don't have a translation for it.

A neat way of seeing how English supports this transcendence of the vernacular "true" is seeing how "true" can be used to provide truth conditions for sentences not in our own language. Consider this acceptable sentence of English:

(1) "Chaque champignon est vénéneux" is true if and only if all plants without chlorophyll cause those who eat them to become ill or die.[10]

A couple of points. First, as I said, (1) is good English. Straightforwardly included in *English* are quoted parts of other languages; one reason for this is so that "true" can conveniently play its transcendental role toward linguistic items conscripted from foreign tongues. Second (and this is especially true for those English speakers *sans* knowledge of French), should someone (perhaps on the basis of knowing (1)), go on to assert:

(2) "Chaque champignon est vénéneux" is true,[11]

this is an ineliminable blind truth-endorsement (what's needed to be said can't be said without "true," in contrast to " 'snow is white', is true") because we can't *say* "Chaque champignon est vénéneux" in *English*.

Conclusion: In contrast to Tarski's reconstructed notion of "truth," our vernacular truth idiom is transcendent in Quine's sense: Its scope includes sentences from any language whatsoever, whether we understand those sentences or not, and whether a translation of those sentences into our own language is even, in principle, possible. Furthermore, this isn't just an oddity we can do without; its transcendence is part of what makes it indispensable for us.

2.3 Criticisms of Tarski-Truth on the Grounds of Its Immanence

Criticisms of the Tarski truth construction that focus on the resulting predicate's immanence have often been raised in the literature. It has been noted, as

---

sarcastic suggestion that perhaps he was unfamiliar with how the word "true" is used. Instead the natural response is to challenge his *epistemic warrant* for his claim. But this shows that blind truth-endorsements of claims we can't state explicitly, and otherwise don't understand, is routine.

[10] I draw this example from Dummett 1999, 279.

[11] Imagine that items like (1) appear in a game based on truth conditions for French sentences. Having read reviews of the game that testify to its accuracy, I can draw conclusions about certain French sentences being true or false without knowing what they mean.

we did in 2.1, that Tarski's construction doesn't indicate how new logical idioms, if added to the object language, are to be handled; new truth-condition clauses must be introduced ab initio—and sometimes without guidance because the new truth conditions can't be straightforwardly modeled on the old ones. So, too, it has been objected that Tarski's approach doesn't supply either a definition of, or even axioms for, a formalized version of truth per se, but only definitions of, or axioms for, an open-ended family of truth predicates true-in-L_1, true-in-L_2, and so on. And, some have added, we can't even be taken to understand these various truth predicates, in particular, understand what makes them *truth* predicates (members of the same family of logical idioms) unless we presuppose the transcendent notion of truth that we have supposedly left behind.[12]

Neither of these concerns has ever struck me as having much force. For whether one feels force from them turns on exactly how one understands Tarski's having "defined" truth. Those who feel there is a philosophically substantial notion—Truth—that Tarski's approach is meant to illuminate will press the idea that our understanding of how all these various formal analogues (true-in-L_1, true-in-L_2, etc.) are supposed to be formal analogues of the same notion requires our presupposed understanding of the vernacular "true," and thus can't illuminate it. If, however, like me (and like Quine, I might add), one focuses on the role of the truth idiom, the need for something that facilitates use/mention divides in the context of blind truth-endorsements, one can simply note that the Tarskian item—assuming, of course, we utilize formal analogues of natural languages instead of those languages themselves—can be seen as designed to fulfill a certain role, and the various truth idioms' capacities to fulfill this role are all that's needed to tie together the various true-in-L_is.

A similar point can be made about the technical concern of how to extend Tarski's approach to new idioms. One needs to keep in mind what the resulting truth predicate is supposed to do, and proceed accordingly. This isn't to say, of course, that "proceeding accordingly" need be straightforward—on the contrary. What it *is* to say is that it will be quite clear whether the result achieves the goal.

However, this last way of responding to the worries I urged us not to take seriously raises an earlier worry about the immanence of Tarski's truth predicate that must be grappled with. This is that the role that the transcendent vernacular "true" is for—blind truth-endorsement—isn't a role (as we've seen) that Tarski's truth predicate, nor (as we'll see), any of the alternative technical devices on the market (e.g., sentential substitutional quantification, infinite conjunctions and disjunctions) can handle: All such approaches are restricted to particular languages, and the vernacular truth predicate isn't.

That the alternative technical approaches can't handle the transcendence of truth might come as a surprise: After all, Tarski's approach is intimately linked to the syntax of each language that a Tarski-style theory of truth is constructed for. Thus it may seem that all we need to do to transcendentalize

---

[12] A more recent version of this rather old objection may be found in Blackburn 1984, 266–67.

truth is to liberate it from syntax. However, things will prove more difficult than we might have expected—not because it's hard to liberate truth from syntax (actually, that's pretty easy)—but because the culprit forcing immanence on Tarski-truth isn't its intimate link with syntax, but Tarski's adequacy condition for a theory of truth: Convention T.

2.4 Making the Immanent Do the Work of the Transcendent: Objections to the Translation Strategy

The claim that immanent truth devices can't handle blind truth-endorsements of sentences from other (natural) languages may strike some philosophers as less than obvious[13] since two ways of responding to the claim come pretty quickly to mind. First, there is the possibility of replacing sentences as the primary truth vehicles with propositions, where the latter are taken to be translanguage items. I'll take up versions of this maneuver in 2.6 and 2.7.

Second, there is the claim that the immanence of a truth idiom is in fact no restriction at all. Each (natural) language has the resources—via translation—to express truth conditions (like (1) in 2.2) for the sentences of any other foreign natural language, and so an immanent truth predicate, wedded to a particular natural language, nevertheless can be used to blindly endorse the truth or falsity of any sentence in any language by means of its translated proxy.[14] There are, broadly speaking, two ways that one might try to support the claim that translation supplies everything that's needed. First, one might claim that—as an empirical matter—it can be shown to our satisfaction that natural languages are rich enough in their vocabulary resources, and in mutually similar ways, so that translations of sentences from any one such language to any other exist. One will then supplement this conclusion by claiming in addition that a truth predicate that can handle blind truth-endorsements of sentences of (other) natural languages suffices: Artificial languages aren't ones that we can even in principle use, and so they can be left aside. Second, one might claim that—on methodological grounds—the very idea of languages, "conceptual schemes" that have sentences inexpressible within ours, proves to be incoherent upon careful inspection.[15]

The second move is a philosophically intricate one presupposing powerful assumptions about the conditions necessary for the interpretation of languages; I must therefore table it for the time being. The supplementary claim about artificial languages is one I also oppose, but can (and do) take up in this book (in chapter 4): I'll claim that although we can't outright desert natural languages, we can (and do) regiment them in ways that require significant

---

[13] For example, Davidson writes: "We credit Tarski with having shown how to make sense of remarks like … 'Everything Aristotle said (in Greek) was false.' " (1996, 269).

[14] Or, to put the suggestion in another way: The evidence for the transcendence of the vernacular truth idiom, given in 2.3, is compatible with that predicate actually being immanent, but supported by rich enough resources for translation that it can handle blind truth-endorsements of sentences of other languages.

[15] E.g., Davidson 1984b.

enough deviations that we need a truth idiom with blind-endorsement powers that go beyond the blind endorsement of sentences only of natural languages.

For the time being, let's consider the claim—presumably to be empirically established—that natural languages are rich enough in their (nonlogical) vocabulary, and in mutually similar ways, for translation to successfully support blind truth-endorsements of sentences in foreign languages. The argument I'll give to oppose this claim relies on empirical facts about scientific theories, some of which are justified elsewhere. I'll try, nevertheless, to make the argument here as self-contained as possible.[16]

Consider some "theoretical" terms of our language—those, say, introduced into our language from particle physics during the rise of the standard model in the 1960s and 1970s. Call English around 1885 "old English," and call English around 1985 "new English." Can all the sentences of new English be translated into old English? No, because the terms for the subatomic particles and their properties depicted in the standard model form a definitionally independent unit: Although they are interdefinable among each other in some cases (e.g., "electron," "charge," "spin," and so on), they aren't definable in terms of the vocabulary of old English. This point is hardly controversial. Recognition of it by philosophers, early on—that is, recognition that one can't define theoretical terms observationally, in terms of instrumentation or, more broadly, in terms of the observational implications of the theory (that theoretical terms are embedded in)—was used polemically against operationism.[17] Exactly similar considerations show that such theoretical terms aren't definable in terms of the ordinary macro-objects we interact with either. But then no translations of such theoretical terms into the language of ordinary life are possible: Translation into a language requires the possibility of definition in the terms of that language.

This might seem to imply, in turn, that the terminology of the standard model that the word "electron" is situated in is utterly detached from the vocabulary of ordinary life. Not so, and for reasons exactly analogous to what happens when new terms are introduced for items that we *can* observe. Suppose a new sort of rock is found. One learns that the presence of this kind of rock can be recognized on the basis of certain "symptoms." In the easiest case, those symptoms may be how that sort of rock looks (perhaps under certain kinds of light) or how it responds to heat, and so on. But things can be more subtle: Our ways of recognizing the presence of this kind of rock may turn on sophisticated chemical tests that show the presence of certain compounds this kind of rock always contains. When we introduce a name for this kind of rock— say, *Saxum novum*—we don't *define* the name in terms of our current theory about the kind of rock (for that theory could turn out wrong) nor in terms of

---

[16] There is another argument against the translation view that turns on the claim that different languages can house different logics, and that two languages obeying the dictates of two different logics often can't be translated to one another. This argument—which I take very seriously—is one I can't defend now.

[17] See Hempel 1965, 128–30.

our current procedures for recognizing the presence of the kind of rock (for those could turn out wrong too); we simply accept—provisionally—these methods of determining the presence of that kind of rock, and in this way (and in this way alone) anchor (loosely) the term to our terms of ordinary life (in which our descriptions of the evidential procedures for determining that *Saxum novum* is present are largely couched).

Things are—in principle, anyway—no different for the "theoretical terms" that refer to items that are far harder for us to get access to (subatomic particles, say). Here too, our usage of such terms (and what they refer to) isn't cemented referentially by theory but by instruments that play the same epistemic role that (sophisticated) observation plays in the case of *Saxum novum*. That is (to put the point philosophically), it's done by forging causal relations (in real time) between individual scientists or groups of scientists and, in this case, instances or groups of such subatomic particles.[18] By courtesy of our belonging to the same linguistic and epistemic community as these scientists, we take ourselves to have the same causal relations to subatomic particles (so that we can refer to them as successfully as scientists do despite our sociological remoteness from the origins, strictly speaking, of the causal relations between these particles and our community of speakers). But should you belong to a linguistic community in which such relations aren't forged by any members of that community, you won't therefore (even by courtesy) have causal relations to these particles, and so the terms in question can't be ones in your language.[19]

2.5 *Making the Immanent Do the Work of the Transcendent: Responding to the Objection Just Made*

It may be felt that even if the considerations just raised against the translation strategy for enabling immanent truth devices to function transcendentally are right, an illicit reliance on the somewhat contingent limitations in nonlogical vocabulary in ordinary language has been (inappropriately) in play. Consider a time (long, long ago) when English lacked a term for Diet Pepsi. Should someone have come across the stuff, even without the term, she could have expressed the sentence

(3) Milk tastes better than Diet Pepsi,

with the sentence

(4) Milk tastes better than *that*,

---

[18] See the articles in Hoddeson et al. 1997 for numerous illustrations.

[19] I've obviously simplified the analysis of the scientific situation *a great deal*. Nevertheless, I'm relying on certain claims that I've tried to establish elsewhere with the detailed argument they deserve: that our "thick epistemic access" to theoretical entities is on a par, epistemically and metaphysically, with our "observational" access to the macro-objects that surround us, and that the role of scientific theory used in conjunction with such thick epistemic access is similar to the kind of generalizations we (implicitly and explicitly) utilize when we sensorily track objects. For how this is supposed to go, see Azzouni 2004c; 2000b, parts I and II; and 1997.

where "that," accompanied by a natural dismissive gesture, refers to a recently scorned glass of Diet Pepsi.

(Some) truth deflationists have notorious problems handling blind truth-endorsements to sentences like (4) with demonstratives. I'm going to assume that such sentences can be managed in a way compatible with BTDist doctrine, and on the basis of this assumption, provide a defense for the translation strategy against the objection raised in 2.4.[20] The defense is this: If we allow that we can very naturally extend the scope of what we can express in a language beyond the particulars of the lexicon of our language, we can't blithely assume that there are sentences that fail to be translatable into our language on the mere grounds that vocabulary items in those sentences fail to be definable in our language. Turning a notion of Hofweber's (2005) to our purposes here, we can say instead that—although ancient Greek doesn't allow for the expression of (3)—since (in the context where there is Diet Pepsi right in front of a person from ancient Greece) she could utter (a translation into Greek of) (4), (3) is "loosely speaker expressible." "Loose speaker expressibility" of (3) requires only the *possibility* of a context in which a speaker can demonstratively indicate Diet Pepsi, and say (4).

As the example of the unexpected presence of Diet Pepsi in ancient Greece makes clear, it simply isn't appropriate to claim that the expressive resources of a *language* extend beyond the resources of its present lexicon all the way to *whatever* is "loosely speaker expressible." It's perfectly bizarre to claim that what amounts to (3) can be expressed by ancient Greeks in ancient Greek by (4) on the grounds that (4) itself *would have been* expressible *had* the ancient Greeks invented Diet Pepsi, or *had* a peculiar accidental chain of events generated a pool of it in front of a thirsty ancient Greek willing to take a chance drinking the unpleasant-looking brown liquid.[21] But, on the other hand, perhaps it's too much to say that the expressive resources of a *language* should be restricted to what Hofweber (2005) calls the "factually speaker

---

[20] Recall that the BTDist, on my view, only wants T-biconditionals to fix (correctly) the truth values of blind truth-endorsements. To do this, one *may* have to supply truth conditions that are not of the T-biconditional form to fix the truth predicate in value appropriately; by rejecting the conceptual fundamentality of the T-biconditionals, the BTDist can be regarded as open to this move. On the other hand, the truth conditions in question can't be ones that violate the neutrality stance of the BTDist. That is, although the BTDist should be willing to allow the possibility of a substantial notion of truth, she shouldn't allow the success of blind truth-endorsements to require a substantial notion. So the resulting truth conditions for demonstrative-laden sentences shouldn't *presuppose* a substantial notion of truth. As I said, I'm going to assume for the sake of argument that a project like this can be carried out. Those who disagree can move on to 2.6, since on their view I'm defending the objection to the translation strategy against a rejoinder they don't take seriously. I say more about the requirement of the BTDist's theory of truth not presupposing a substantial theory of truth in 2.7, and more about demonstrative-laden sentences and how to handle them in 3.5.

[21] We can't, that is, assume that a group of people speaking a language can use gestures toward some object or kind of object to express truths on the mere grounds that *somewhere* in space and time that object *exists*. It has to be reasonable, in addition, to claim that the group of people in question have epistemic access *to* that object, or to instances of its kind.

expressive"—where that notion is gotten from "loose speaker expressibility" by cutting down the contexts available to only those the speakers in question (ancient Greeks, in this case) are (were) actually *in*. This may be regarded as too restrictive by some because there are contexts that, in a perfectly robust sense, ancient Greeks *could have* been in: e.g., they could have gestured at a particular animal with an unusually colored mane that they would have seen if they'd traveled only a few miles farther than they did.

I won't try to adjudicate exactly what sort of extension—via contexts—we should allow as legitimate in the expressive powers of our languages beyond their actual lexicons at a time. All I need to sustain the objection to the translation strategy is that it be accepted that an extension that allows, for example, background cosmic radiation among those items ancient Greek can express truths about, is an extension that goes way too far: The theoretical knowledge plus technical know-how that's required to gain access to such a thing (and other things we currently can talk about) is beyond—in a perfectly clear sense—what ancient Greek speakers have access to by way of contexts that can be used to augment the expressive powers of their language.[22]

2.6 Making the Immanent Do the Work of the Transcendent: Translanguage Propositions

It should be clear that the considerations raised in 2.4 do as much damage to *one* version of the propositional strategy as they do to the more direct translation strategy. For imagine that one presumes that propositions are translanguage items that sentences of particular languages express, and that are supposed to be the actual vehicles to which we attribute truth and falsity. Nevertheless, on one version of the view, every such proposition so understood is one expressible in any natural language, even if speakers—like the ancient Greeks—have no (even in principle) resources for expressing these propositions. It's hard, if the considerations raised in 2.4 are right, to see this version of the propositional strategy as anything more than an ontological act of fiat: an a priori imposition of translational adequacy on natural languages solely for the purpose of supplying enough resources so that an immanent truth predicate can function transcendentally.

A more reasonable variant looks possible, however. This is to allow languages to be expressively variable in what propositions they can express, but to build a Tarski-style theory of truth for a language not solely on the propositions expressible in that specific language, but over the entire domain

---

[22] There is, of course, yet another way of arguing that all (natural) languages have the same expressive resources, and therefore, are (in principle, anyway) mutually translatable into one another. This is to claim (as Fodor 1975 does) that we can acquire natural languages only because we have an innate species-specific language of thought that the acquisition of a natural language is actually the translation of such a language into. The view is a controversial one, in any case; although I think it's wrong, it raises substantial and interesting issues that I can't get into now.

of propositions that occur in any language whatsoever. If this project could be executed, it would formalize a transcendent (Tarskian) truth predicate.

But there are great obstacles to the suggestion that this option is even *cogent*. First, the domain of propositions in question is unlikely to be a set, and this may pose problems for a *Tarskian* approach. Second, the certainty that different languages obey different grammatical principles, plus the open-ended number of nonlogical vocabulary items (each one of which will require a separate truth clause), will make such a theory of truth deeply intractable—in the sense that the Tarskian truth clauses for such a theory won't be ax-iomatizable. Notice what the point isn't: that the resulting Tarskian theory is incomplete. Rather, the point is that in order for the truth predicate to be exercised in its transcendental role, we must have access to *all* the truth clauses, and wedded as such clauses are to the foreign syntax, we must therefore have access to those clauses as well. This is too much to demand of a speaker trapped (as speakers are) within the grammatical confines of the languages they know.

It might be hoped—although on what grounds is hard to say—that grammatical differences in such languages are a superficial matter: The propo-sitions that sentences from any (natural) language express are structured simi-larly so that a tractable set of construction axioms is applicable to all of them. Still, the base clauses must be given explicitly for the nonlogical vocabulary that differs from language to language. And this exposes the essential oddness of this suggestion: The truth predicate in question is supposed to be one that's ex-pressible in a particular language. And yet, the rules governing its use essentially depend on linguistic resources that the speakers in that language don't—by assumption—have. Of course, in general, Tarskian metalanguages have re-sources that go strictly beyond those of the object language the truth predicate is crafted for. But the assumption must be made that the speaker's notion of truth is hierarchically structured along with *his* language, so that his language has the resources not just of the object language that the truth predicate is for, but also of whatever is needed for his truth endorsements. An assumption like that, in a context in which the truth predicate in question can play a tran-scendental role, amounts to including among the resources of the language **L** of the speaker that the "immanent" truth predicate belongs to, those in addition that allow the construction of *all* the propositions (that by assumption *aren't* expressible in **L**) that the speaker can blindly truth-endorse. But then this suggestion reduces to the one dismissed at the beginning of this section.

2.7 The Drawbacks of Non-Tarskian Approaches to Truth That Are Also Immanent: More on Propositional Quantification

The failure to transcend a particular language is a problem that besets not just Tarskian approaches to truth, but *any* approach that restricts truth to a par-ticular language. In 2.8 and 2.9, I'll focus on the sentential substitutional approach to truth, and generalizations of such; but I should say something more about why nominal-quantifier approaches are being left aside.[23] The reason stems from considerations raised at the end of 2.6. Assume (again) a domain of propositions, **D**, where **D** contains every proposition expressible in any (natural) language but the (full) contents of which aren't expressible by (all) the sentences of any one language. How are the truth values of blind truth-endorsements (from within the context of any particular language) to be determined via **D**?

In principle, actually, it may seem that very little is needed to fix these truth values—hardly the full Tarskian apparatus. Call the set of T-biconditionals for all the propositions in **D** the "Minimal Theory of Truth" (MT). It should be clear that in these circumstances MT suffices for blind truth-endorsements from any particular language to any other.[24] Given an assignment of truth values to the propositions of **D**, every blind truth-endorsement is fixed in truth value as well.[25]

Notice that we can't *prove*, using MT, that MT is adequate for blind truth-endorsements. But that's not a requirement on a theory of truth—that it be able to prove its own adequacy for certain purposes. All that's required is that it *be adequate* for those purposes.[26]

So I find myself disagreeing with most of the arguments in the literature that have been put forth against MT. Nevertheless, there is—I think—a fatal objection to MT; this is that it faces exactly the same problem we saw the Tarskian approach face at the end of the last section. Consider a notion N governed by an infinite set of axioms A that are recursively enumerable—this is, presumably, the model according to which we understand how MT governs

---

[23] Versions of these are fairly popular among philosophers. See, e.g., Richard 1996 or Soames 1999.

[24] I'm setting aside (until chapter 4) issues about "self reference": in this specific case, blind truth-endorsements from speakers of two different languages of the statements of one another that can't, as result of what they claim about each other, be "grounded."

[25] David (1994, 71) argues that blind truth-endorsements can't be handled by MT. (Actually, he argues that they're fatal to a generalization that yields only MT as instances.) His argument turns on requiring a theory of truth to yield a method of eliminating the term "is true" from blind truth-endorsements. That's not a requirement here. Indeed, it's hard to see—if we keep our eyes firmly on the role of the truth predicate being solely to fix blind truth-endorsements (appropriately) in truth value—why it *should be* a requirement.

[26] It may seem surprising that MT is adequate for blind truth-endorsements. Perhaps we've overlooked something: Maybe a theory of truth must do more vis-à-vis blind truth-endorsements than simply fix their truth values. *But what?* The only reasonable candidate would be licensing inferences from truth claims to other truth claims. Leaving aside examples of such that rely on already controverted substantial semantic principles (such as McGee's "The Pope uttered a true conjunction, one conjunct of which was 'To be is to be the value of a variable; conclusion: To be is to be the value of a variable'"), it's hard to think of *any* that aren't already licensed by the ordinary logic of quantifiers (ranging over propositions, in this case)—e.g.: Everything John uttered is true; John uttered Sam's conjecture; Sam's conjecture is that cats come in many colors; cats come in many colors. None of this requires more than MT. Notice, by the way, that MT doesn't fix the extension of the truth predicate in the sense that it can be *proven* on the basis of MT that any two "truth predicates" will be coextensive. (Also note that I'm assuming that the move to propositions solves problems with ambiguity and demonstratives.)

truth: by an infinite set of T-biconditionals, one for each proposition in **D**.[27] However, in practice, any proof about N is graspable on the basis of A: That is, there is a decision procedure for determining of any purported axiom used in this proof whether it appears in A. But nothing similar is true of MT. This is because, by assumption, for any language L, there are instances of MT that involve propositions not expressible in L, and so those instances of MT won't make sense to the speakers of L. In what sense, therefore, can those instances of MT even be *relevant to* the use of the truth predicate as employed by speakers of L?[28]

Here's another way to make the point. Imagine two truth predicates, T and T*. Only the sentences of our language fall under the extension and anti-extensions of T. The extension and anti-extension of T*, on the other hand, includes all propositions, regardless of whether they are expressible in our language. What indicates that *our* truth predicate is T* rather than T? One suggestion that would do what's needed is that truth is a substantial property, and that our truth predicate picks out that property. Then, by virtue of the property *truth* that the propositions not expressible in our language also partake in, such propositions will also fall under *our* truth predicate (or under its anti-extension). Unfortunately, any BTDist who wants to remain meta-physically neutral can't go this route.[29]

The problem with MT, on the one hand, and the problem with a Tarskian truth theory, on the other, are the same: that sentences (or propositions) on the right side of the T-biconditionals are *used*, not mentioned. If they are the

---

[27] See Leeds (1978, 122) and Halbach (1999, 4). Horwich (1998, 34–35) draws an analogy with "the idea that our conception of number is determined by the disposition to accept Peano's Axioms, including infinitely many instance[s] of the induction schema." Incidentally, this suggestion will *not* do for our concepts, specifically, our concept of number. See 7.11.

[28] Horwich (1998, 128) writes, by way of an attempt to meet the objection that the minimalist approach can't handle truth endorsements of untranslatable sentences: "The minimalist thesis is that the meaning of 'true' is constituted by our disposition to accept those instances of the truth schemata that we *can* formulate." Let's grant this. " '@4^^&*' is true if and only if @4^^&*," is meaningless to English speakers, *despite* our disposition to accept those instances of the truth schemata that we *can* formulate. Why, therefore, assume that the meaning of "true" so "constituted" extends one *single* proposition beyond those expressible *only* in our own language?

Notice that what we need, and what I'm assuming that we need, is some reason to think that MT—*all* of MT—governs our truth predicate; because without that, the truth values of our blind endorsements will not turn out correctly. In turn, something about our verbal practices must reveal how *all* of MT is involved. But precisely that is what the listlike quality of MT, coupled with the use (on the right side of the biconditionals) of statements not in the language of the speakers, prevents.

[29] The proponent of MT can, of course, regard the two truth predicates T and T* as exemplifying "the same notion" by virtue of their having the job of semantic ascent and descent. But this isn't enough to enable the claim that those restricted to a particular language nevertheless *have* a truth predicate that governs propositions not expressible in that language.

I should add that, even granting such a property *truth*, there would still remain an issue of exactly *how* our truth predicate was capturing the property of truth—since it clearly couldn't be doing so by virtue of T-biconditionals that speakers grasp: We can't just *grant* that the extension of terms are given by properties that (conveniently) hold all of the objects that we need those terms to refer to, can we?

items that fix the extension of the truth idiom, and if, therefore, they can't include in their company sentences not in the language **L**, nor propositions not expressible by sentences in the language **L**, it's hard to see how blind truth-endorsements of statements not in **L** are possible. This suggests that any approach which accepts Convention T as an adequacy condition of a theory of truth, including ones that use sentential substitutional quantification, faces exactly the same problem. And indeed, as we'll see, it does.

2.8 The Drawbacks of Non-Tarskian—But Still Immanent—Approaches to Truth: Infinite Disjunctions and Conjunctions

We turn now to the consideration of devices that facilitate blind endorse-ments[30] by means of some version of a desertion of nominal quantification. The first strategy I'll consider is the *replacement* of the truth predicate by devices that *look like* substitutional quantifiers (($\mathbf{E}x$) and ($\mathbf{A}x$)). "Something written on the blackboard is true" is thus regimented in this idiom as: ($\mathbf{E}x$)(Written-on-the-blackboard"x" & x).[31]

I say "look like" because one popular way of construing such items (recall 1.3), and a way commonly seen as fitting best with the overall deflationist strategy, is as abbreviations for infinite disjunctions and conjunctions of sen-tences of the language they appear in. So, for example, ($\mathbf{E}x$)(Written-on-the-blackboard"x" & x) is short for: Written-on-the-blackboard"A" and A or Written-on-the-blackboard"B" and B . . . , and so on, for every sentence in **L**.[32]

Unfortunately, given the need for a device that facilitates blind en-dorsements of sentences not in **L**, this won't do: A replacement of the idiom

---

[30] Let's not get tripped up by terminology. As understood, "blind truth-endorsement" needn't involve the idiom "true." So I'll henceforth call it "blind endorsement."

[31] This *looks like* quantification into quotation marks, but the forthcoming (abbreviatory) interpretation of it shows that it isn't.

[32] Leeds 1978; Williams 1988, 424; Simmons 1999. It's not assumed, as I understand this strategy, that the truth predicate stands stead for devices of infinite conjunction and disjunction. Halbach 1999 is entirely directed toward attacking that view; but I'm not convinced that defla-tionists drawn to this strategy ever claimed *that*. Rather (I should think) the idea is that the truth *predicate* is to be replaced by tools available via devices of infinite conjunction and disjunction. It also seems that the species of infinite conjunction and disjunction that suffice for this purpose is very simple: *All* the sentences of the language **L** (listed, let's say, in some manner) will do. Since, as just mentioned, the best way to see this strategy is as an eliminativist one—the truth predicate is replaced by something else that can facilitate blind endorsements—the need to consider (ex-plicitly) a truth predicate, or the T-biconditionals, lapses because the need for a device that crosses use/mention divides itself lapses. (Each clause of the infinite conjunction or disjunction utilized to blindly endorse both uses a sentence of the language, and via quotes, mentions it.) This doesn't mean, of course, that a truth predicate couldn't be *defined* using these devices, and one that provably obeys all the T-biconditionals of the language **L**. Of course one can. But on a view that the aim is to find an idiom—truth predicate or something else—that facilitates blind endorse-ments, there is no point to defining "truth" once such devices are available: They can do the job themselves.

of truth with abbreviatory devices for infinite conjunctions and disjunctions can't provide a transcendental truth idiom for exactly the reasons already considered in 2.4, 2.5 and 2.6: Such conjunctions and disjunctions are, of course, drawn only from the language L, and thus don't contain any of the sentences of a foreign language that a speaker of L may blindly endorse.

2.9 Purported Problems with Using Sentences in the Substitution Class of Sentential-Role Substitutional Quantifiers

Let's turn, therefore, to strategies which explicitly introduce substitutional quantifiers ((Ex) and (Ax)), which I'll call sentential-role substitutional quantifiers (hereafter, SRS-quantifiers).[33] These are, unlike the abbreviations for disjunctions and conjunctions of the last section, genuine quantifiers that have as their substitution class the sentences of L.[34] One construal of their semantics (roughly) is this: $((Ex)S(x))$ if there is a sentence s in L (in the substitution class of the quantifiers) and if $S(s)$, and similarly for the universal quantifier.[35]

Before exploring any further the virtues (and vices) of this strategy for supplying a device to facilitate blind endorsement, I should discuss what has sometimes been taken as a problem for sentential substitutional approaches, and thus as a motivation for opting for (the previous rejected nominal) quantification over propositions instead. Start with the obvious point that the sentences of natural languages are ambiguous. Philosophers in favor of the general strategy of quantifying over propositions use this general ambiguity against those who prefer sentences as truth vehicles—for example, proponents of SRS-quantification. The worry is this: Sentences of the same physical form, and in the same language, can nevertheless "mean" different things, and furthermore, sentences of the same physical form can "mean" different things in different languages. Consequently, if "sentences" are to be truth bearers, they can't be individuated only in terms of their (sheer) physical appearance.

If we want to insist that something like sentences-types (and not sentence-tokens) are (consistent) truth bearers, then we must either regiment ordinary language free of ambiguity, and only allow thus regimented sentences to be

---

[33] The reasons for this—somewhat wordy—nomenclature will emerge shortly.

[34] For details, see Kripke 1976.

[35] Why not: $((Ex)S(x))$ *is true* if there is a sentence s in L (in the substitution class of the quantifiers) and if $S(s)$ *is true*? Because if L is part of the language in which the semantics is given, or if suitable proxies for the sentences of L are available (as in the standard Tarskian approach), the "truth conditions" of these quantifiers can be expressed in the metalanguage directly in this fashion, bypassing the word "true"; and I'm indicating that explicitly by the formulation I've given in the text. I should add that if the considerations pressed in chapter 1 are right—that all uses of "true" in truth-conditional semantics involve only its use/mention navigating role—then the appearance of the word "true" in the truth conditions of SRS-quantifiers is benign in any case; it's operating as a placeholder in a locution that can be explicitly replaced by locutions using *whatever* devices do the job of facilitating blind endorsement in the metalanguage.

admissible truth vehicles; or we must individuate "sentences" differently than physical distinctions in word type and syntactic structure individuate them. In the latter case, in the words of Richard (1996, 438): "the substitutionalist will have to individuate sentences in terms of an intra-linguistic relation of having *the same sense as.*" Richard seems to think that the closer the substitutionalist is pushed toward individuating sentences in terms of propositions that such sentences must be taken to express, the closer the substitutionalist is to nominal quantification.[36]

(For the record) I'm unconvinced that presence of syntactic and lexical ambiguity in ordinary-language sentences requires one to move from sentences to propositions (as the primary truth bearers) that such sentences must be taken to express. But let me grant the suggestion, for conceding it actually has no impact whatsoever on the question of whether the *quantification* utilized in theories of truth should be nominal or sentential.

Recall that the problem that nominal quantification faced (in 2.6 and 2.7), when coupled with the need to fix the extension of the truth predicate by means of T-biconditionals, was that the resulting theory of truth couldn't be transcendentalized without, in effect, importing the entire expressive resources of foreign languages into the language housing the truth predicate. Consequently, I'm now exploring approaches to truth that eschew nominal quantification, in particular the alternative of substitutional quantification into *sentential positions*, in the hope that so altering quantification (in the context of a theory of truth) may get around the problem with immanent idioms that can do no transcendental work.

In this case, the substitution-instances that stand stead for variables in sentential position, are sentences. But nothing, per se, in the SRS approach forces one to deny such sentences are to be individuated by the propositions they express; indeed, nothing about the SRS approach prevents the placing of propositions (themselves) in the substitution class, indexing those propositions to *sentences* that express them, and allowing the substitution of such sentences for variables in sentential position to be linked to the truth conditions of the SRS-quantifiers.

Indeed, this looks unproblematic as long as there are sufficient sentences in L so indexed to the propositions themselves that so express those propositions. Ambiguity itself raises no issue, for sentences-on-a-sense, may satisfy truth conditions for a SRS-quantifier:

---

[36] Richard (1996) doesn't quite *say* this, but after raising these, and other, issues about whether sentences can appear in the substitution class of a substitutional quantifier, as opposed to something akin to propositions, Richard moves directly on to a proposal couched in terms of *nominal* quantification over propositions. For the sake of argument, I'm going to concede the need to individuate sentences by (say) the propositions they express, but resist the move to *nominal* quantification nevertheless. I should add that Richard's discussion raises issues not only about ambiguity but also about demonstratives. The latter offers a more subtle problem for sentential substitutional approaches—and one, arguably, they can't meet. The forthcoming AU-quantifier approach, however, *can* handle demonstrative-laden sentences. See 3.5 for a discussion of this.

(5) $(\mathbf{E}x)(\text{WRITTEN-ON-THE-BLACKBOARD}(x) \ \& \ x)^{37}$,

is true if, for example, "Some maps are drawn in blue ink," is written on the blackboard, that is

(6) WRITTEN-ON-THE-BLACKBOARD (Some maps are drawn in blue ink),

is true, and the token of "maps" that appears on the blackboard belongs to the type MAP, all of whose instances refer to the physical items, and not to the mathematical items; and (finally) that some maps—in this sense—*are* drawn in blue ink (which they are).

Some may worry that sentences, in this sense, aren't the items written down on blackboards. What's written down on blackboards are (tokens of) purely physical items. But this should be denied: If sentences that make sense are items that we routinely describe as appearing on blackboards, and in newspapers (and we do), then such items should be individuated—on this view—propositionally; and that means that the tokens of them that are to be substituted for variables in sentences like (5) must be similarly individuated on the basis of their sense.

2.10 Possible Drawbacks of Non-Tarskian—But Still Immanent—Approaches to Truth: Sentential-Role Substitutional Quantifiers

However, if the substitution class of the SRS-quantifiers, however refined, is restricted to the sentences of L, we face the same problem that nominal quantification faced in earlier sections of this chapter with respect to the blind endorsement of statements from other languages. It might be thought, though, that since the semantics of the SRS-quantifiers rely on substitution classes, one can open substitution classes of the quantifiers to sentences from other languages. If I blindly endorse the utterances of a French speaker, my claim "Pierre said something true" is evaluated, according to this version of the SRS approach, via a substitution class containing both the sentences of French and those of English. We don't, however, want to understand this suggestion in a way that *obviously* faces a version of the objection raised to the various nominal quantificational strategies considered in 2.5 and 2.6. We don't, that is, want blind endorsements of sentences from foreign languages at the cost of importing into our own language all the resources needed to express these things. But if the sentences of French are included in the substitution class of the quantifiers, then, as substitutional quantification is sometimes understood, expression of those sentences becomes available in the *extension* of L *that's now taken to be the language of the speaker*. If I can say:

(7) $(\mathbf{E}x)(\text{PIERRE-SAID}(x) \ \& \ x)$,

---

[37] "WRITTEN-ON-THE-BLACKBOARD" must be understood on this approach as a sentence-operator. I discuss the problem this gives rise to in 2.10 and 2.11.

then I can say,

(8) PIERRE-SAID(Chaque champignon est vénéneux) & Chaque champignon est vénéneux,

if "Chaque champignon est vénéneux" is what Pierre said, and, in addition, I think it's true. But blind endorsement of the sentences of foreign languages should not yield the capacity to express foreign sentences so cheaply.

We, should, therefore allow the substitution class of the quantifiers to contain sentences from other languages *without* those sentences being allowed (as a result) to be among the sentences of the speaker's language—without it being legitimate (for the speaker) to use them: The truth conditions, that is, of SRS-quantifiers must take account of sentences well beyond those of the speaker who is presumed to know only L. In other words, because the truth conditions are concerned with sentences in the domain, and not those merely in L, SRS-quantifiers are more akin to objectual quantifiers than to substitutional ones because their truth conditions turn—strictly speaking—not on sentences that can be substituted in place of the variables but rather on the variables taking on the role of bearing truth values according to what sentences (and their truth values) are in the domain.[38]

Richard (1996, 449–50) treats this suggestion harshly:

> It makes English semantics include the semantics of every language there is, was, or will be: on this view, English semantics involves reference to the truth-conditions of *Katya said that* S where S is an arbitrary sentence of an arbitrary language. Here S is used, not mentioned. And with propositional quantification into arbitrary sentential position, as in 'For some *p*, *p*', the semantics involves reference to the truth-conditions of arbitrary sentences of arbitrary languages.[39]

We might hope to get around this (I once so hoped) by noting that the SRSist needs only the *sentences* of foreign languages in the substitution class of her quantifiers. So the amalgam of English and French that's currently being contemplated as the substitution class for the SRS-quantifiers, one might

---

[38] The idea is that, when formalized, the truth conditions for quantified sentences would relate to substitution-instances not necessarily in the language L that the speaker can be legitimately taken to know, but with respect to "language extensions" of L. If this is to be regarded as acceptable, one may wonder why a view that takes the operation of MT to succeed because of the existence of T-biconditionals (recall 2.7) even in foreign languages, doesn't succeed on the same grounds. The reason is this: On the current suggestion, genuine quantification is taking place. The speaker doesn't *use* the sentences in the L-extension; rather, facts about the language extension are truth-conditionally linked to quantifications the speaker *does* use. The analogous move in support of MT would be to help oneself to a property of truth that's linked to the use of MT—but this move, as we've seen, deserts the neutrality stance of the BTDist. (And, besides, there are good reasons to deny the existence of any such "property.")

[39] Richard's specific concern is with the substitutionalist construal of belief contexts. But the objection—if sustainable—looks as if it might then apply to SRS approaches to truth as well.

think, need only be a literal pooling of *whole* sentences together, not a pooling of the vocabulary resources of such languages, and recursive constructions from such.

Consider, though, (7), here repeated:

(7) $(\mathbf{E}x)(\text{PIERRE-SAID}(x) \& x)$.

This will be true presuming, let's say, that Pierre said (and only said) "Chaque champignon est vénéneux"; that is, it will be true if (8), here repeated, is true in the appropriate language extension:

(8) PIERRE-SAID (Chaque champignon est vénéneux) & Chaque champignon est vénéneux.

But now we face problems since, as Richard notes, the appearances of "Chaque champignon est vénéneux," are matters of use, not mention. In order to have the quantifier manage the cross-referencing we need, and to appear in sentential position, thus, "PIERRE-SAID" must be treated as a sentential operator the semantics of which must take account of a sentence not in English. One might hope to finesse this (a little) by treating the operator "PIERRE-SAID" as operating on sentences to be construed purely in a syntactic sense; but this is ruled out by our need to individuate sentences more finely than that, by (as I conceded) having to use the propositions expressed by such sentences as what individuates them. Still, it could be argued that the semantics of French isn't required by the semantics of the operator "PIERRE-SAID"— all that's required are the individuation conditions for the sentences-on-a-sense of French, and the results of these individuation conditions (that is, *how* sentences-on-a-sense are individuated in French) can be assumed by the operator "PIERRE-SAID" without the principles of such conditions being (literally) incorporated into the semantics of English.[40]

The lessons drawn about the PIERRE-SAID operator apply to "&." Consider the truth conditions for "&": $(A \& B)$ is true if and only if A is true and B is true.[41] This recursive characterization decides the truth or falsity of $(A \& B)$ by decomposing it into two questions: one about the truth or falsity of A and one about the truth or falsity of B. If A (and B) are sentences of English, then (possibly other) semantic clauses of English apply to them. If not (if one or the other is French) then the answers asked for are given by the semantic rules of French. This requires, of course, that the truth predicate—if present and used—that participates in these truth conditions is

---

[40] That is, the semantic conditions for the PIERRE-SAID operator recursively pass the buck, if needed, to semantic rules in French; but this doesn't require including those rules among the semantic rules of English. Obviously, this strategy is quite sensitive to the operator involved; some operators in English might require more details about foreign sentences that in turn require their meshing with the semantic rules of the language those foreign sentences belong to. In that case, recursively passing the buck may not be straightforward. See the last paragraph of this section.

[41] Recall note 35.

a transcendental one. And of course it would be if we could get our hands on a transcendental truth idiom in the first place![42]

One last point about this. Everyone knows that nominal objectual quantification into non-extensional contexts *can* be troublesome.[43] The issues just raised about the SRS approach are exactly of the same sort: If the sentential contexts are extensional, there is—of course—no problem. Otherwise, whether or not there is a problem depends on the details of the specific nature of the non-extensional context. So there are no special (and distressing) surprises that SRS "quantifying in" gives rise to.

2.11 The Drawbacks of Non-Tarskian—But Still Immanent—Approaches to Truth: Sentential-Role Substitutional Quantifiers (Continued)

Even if sentential-role substitutional quantifiers escape the problems raised in 2.10, there is still another problem: In replacing the truth *predicate* with SRS-quantifiers, anaphora between truth *endorsements* and nominal contexts has been blocked. As we've seen, to handle cross-referencing between what John says and the truth of what John says, SRS-quantification requires treating "John-said," as a *sentence* operator rather than as a predicate: All occurrences of the variable "x" in "$(Ex)(\text{John-said}(x) \;\&\; x)$" must be prosententialized. Transforming all apparently predicational contexts into sentential ones enables us to handle blind endorsements, regardless of the description used to pick out the set of sentences one wants to endorse the truth or falsity of. But there is a problem if one wants to combine descriptions of sentences (and endorsements of those sentences) with descriptions of other objects. For example, consider "Any sentence which is purple is true if and only if some rock is also purple." The natural way to construe this is: "$(x)(Px \Rightarrow x) \Leftrightarrow (\exists y)(Ry \;\&\; Py)$"; but to capture this using SRS-quantifiers requires that we treat the first occurrence of "P" as a sentential operator, and the second as a predicate. And this points to the fact—rather overwhelmingly obvious about ordinary language—that

---

[42] The semantic rules apply, therefore, not just to sentences of English, but to sentences *tout court*. This is a deviation from the normal understanding of the semantic rules of a formal language, which are taken to apply only to the sentences of that language. But one might ask, if the semantic rules of English can apply to sentences other than those of English, why English speakers don't recognize "John is running and chaque champignon est vénéneux," to be both grammatical and sensical (provided, of course, they are told that the second clause is a sentence). A possible answer is that the rules of grammar that they have internalized don't allow them to recognize "chaque champignon est vénéneux," to be a sentence, even if they're told it is. And so, the grammatical rules don't recognize "John is running and chaque champignon est vénéneux," as grammatical either (which, of course, it's not). But this can be conceded without, as a result, conceding that the semantic rules for "&" require that, if the principle connective of something is "&," the two things that the &-rule is applied to must be items licensed as sentences of English by the grammatical rules of English. All the &-rule need require is that the two items flanking "&" be *sentences*—a condition that "John is running and chaque champignon est vénéneux" satisfies.

[43] See, e.g., the classic Quine 1975a, 1975b, 1980b, and the numerous responses made to them.

we can (in principle) apply any description (that we apply to objects of any sort) to sentences *too*, and go on to endorse or deny the sentences fitting that description.

The only other option (if we want to keep to an SRS approach) is to introduce quote-names and quantify *into* these. That is, we treat descriptive contexts as predicational, but tuck within them (technically, within the "quote-names" appearing within such predicational contexts) a sentential context that sentential quantifiers can bind variables within. If "P" stands for "is on the blackboard in room 201," then "$(\mathbf{E}p)(P'p' \,\&\, p)$" is the claim that some sentence on that blackboard is true; and its truth-value status depends on the results of substitutions (of the substitution-instances of the SRS-quantifier) for all its variables within and without quotes.[44]

Quantifying into quotes is supposed to circumvent the obstacle of directly linking nominal devices (e.g., names) with sentences (i.e., those very sentences—or variables standing for them) via SRS-quantifiers. Unfortunately, allowing quantifiers to bind prosentences inside and outside quotation isn't an ideal blend of use and mention with respect to singular-term blind endorsements, if we allow ourselves the use of ordinary (nonquotational) names as well as quantifiers that bind *their* contexts. Consider the context: P'John is running'; and consider the identity: 'John is running' $= (1)$. We can directly draw the conclusion: P(1). But the connections of this inference to quantifiers must run strangely duplicate: We need *both* sentential quantifiers that bind (sentential) variables within the context of "P'—'," and nominal quantifiers that bind the (nominal) variables within the context of "P—," a different matter entirely. Furthermore, we need axioms that allow movement from one sort of quantifier to another—when appropriate. We can then state similar generalizations with either kind of variable, and choose an appropriate form—using either nominal or SRS-quantifiers—depending on which inferences we intend to use the generalization in. But, although this can be done, it's awkward.

Let's admit it: What's needed—what's become *obviously* needed, given the problem just described facing the SRS approach to quantifiers, and given the problems nominal quantifier approaches faced (in 2.6 and 2.7)—are quantifiers that *simultaneously* allow quantification into nominal *and* sentential contexts; and why not just directly introduce the things?

2.12 Concluding Remarks

One lesson that I hope has been established in this chapter is that the ills that beset theories of truth are encapsulated (literally) in T-biconditionals: that in describing sentences and simultaneously endorsing them, one needs (but doesn't seem to have) a device that can simultaneously quantify into sentential and nominal positions. I show, in the next chapter, what such a device is like.

---

[44] See Grover and Belnap 1992 and Grover et al. 1975.

But there's a second lesson that started to become clear in 2.9. This is that once we have another way of blindly endorsing sentences, we can desert Convention T. I (without being explicit about it) considered that very move when considering substitution classes for the SRS-quantifiers that involved more than the sentences of L. For in such a case, our blind endorsements of statements wouldn't be governed by T-biconditionals since those sentences aren't in our language. Instead, the blind endorsements would be directly governed by the logic of the SRS-quantifiers.

This is unwelcome news to those deflationist who continue to insist that the conceptual foundations of our understanding of truth issue straight from the T-biconditionals. But it isn't bad news for deflationism generally. Indeed, that deflationist who has grasped the import of chapter 1—that the truth predicate's role (if it's only a deflated one) can be captured by ana-phorically unrestricted pronouns—will recognize that quantifiers (of some sort) were always doing all the hard work anyway. If there were no need to navigate between use and mention, then only that work of a quantifier (of some sort) would be left. Again, this is unwelcome news to those who take Convention T as an adequacy condition to be placed on any theory of truth; but (again) it's not unwelcome news for someone who takes the function of facilitating blind endorsement to be the adequacy condition to be placed on any truth idiom—whether Convention T is involved or not.

So that brings us to the topic of the next chapter: the introduction of a kind of quantifier, that itself can quantify into both sentential and nominal positions, that can do all the work we need to blindly endorse statements, and further, which isn't governed by Convention T.

3

Anaphorically Unrestricted Quantifiers

3.1 Introduction

It's now time to formalize the anaphorically unrestricted pronouns first introduced in 1.4. The result is a family of logical systems very close in their metalogical properties to those of the first-order predicate calculus, properties that are well understood. I offer two approaches to anaphorically unrestricted quantifiers (AU-quantifiers). The first imitates standard approaches to truth by augmenting an already given language L_0 with AU-quantifiers. Because it allows languages strictly larger than L_0 itself as domains of the quantifiers, blind endorsement of sentences not in L_0 is possible. Thus the ability of the transcendent truth predicate of ordinary language to blindly endorse foreign sentences can be captured by AU-quantification. The second approach to AU-quantification provides a slight generalization of the first which may be useful in certain circumstances.

3.2 describes the language in which AU-quantification lives, and gives its model theory. 3.3 supplies the proof theory. A completeness theorem linking the two exists but won't be given in this book.[1] 3.4 provides a version of AU-quantification that may be valuable for certain special purposes. 3.5, finally, distinguishes the AU approach from substitutional approaches by comparing the abilities of the two approaches to handle sentences with demonstratives.

---

[1] It can be found in the appendix to Azzouni 2001.

3.2 L_1-Languages and Their Model Theory

L_1-languages

Definition 1. An L_0-language (indicated by the variable "L_0") is a standard interpreted first-order language. (I assume the presence in L_0 of unspecified nonlogical constants and predicates.) An L_1-language is an augmentation of an L_0-language by the addition of L_1-variables, "p," "q," ..., the quantifier "(Σp)," and a set Υ of predicates and names which contain "$=$."[2] An L_1-term is either a constant of Υ or an L_1-variable or a sentence of L_0.

Every L_1-language is an L_1-language with reference to a (sometimes implicit) L_0-language.

Definition 2. The well-formed formulas of an L_1-language are so only by the following clauses:

(WF1) If ϕ is a sentence of L_0, then it's a wff of L_1.

(WF2) Any L_1-variable is a wff of L_1.

(WF3) If P is an n-place predicate of Υ, and $\tau_1, \ldots, \tau_n$, are L_1-terms, then $P[\tau_1] \ldots [\tau_n]$ is a wff of L_1.[3]

(WF4) If ϕ and ψ are wffs of L_1 then $\neg\phi$ and $(\phi \,\&\, \psi)$ are wffs of L_1 too.

(WF5) If ϕ is a wff of L_1, and α an L_1-variable, then $(\Sigma\alpha)\phi$ is a wff too.

Definition 3. Sentences of L_1 are well-formed formulas of L_1 without free variables.

By (WF1) and Definition 3, notice, the sentences of L_0 are also sentences of L_1; but those very same sentences, by Definition 1, are *terms*. They thus function both as sentences in L_1 and as (canonical) names for themselves, depending on which context (see below) they appear in.

Examples of sentences of L_1. (i) $(p)(Q[p] \Rightarrow p)$, (ii) $(\Sigma p)p$,
(iii) $(p)(q)(R[p][q] \Rightarrow (p \Rightarrow q))$, (iv) $(\Sigma p)R[p][(\forall x)Ax]$, and (v) $(\Sigma p)(p \,\&\, (\forall x)Ax)$ are all sentences of L_1.[4]

Nominal Contexts and Sentential Contexts

Presence of a sentence of L_0, or of an L_1-variable, within the scope of a predicate symbol (between square brackets), is *presence in a nominal context*; all other contexts for such terms are *sentential*. When a sentence of L_0 is present in a nominal context, I describe it as a *canonical name* (for itself).

---

[2] The system can be generalized without difficulty to include functions. I forgo this.

[3] Following custom, I write "$=$" between terms rather than preceding them, but I still use square brackets: "a = b" appears here as: $[a] - [b]$.

[4] I assume "R" and "Q" are in Υ, that standard definitions of the other quantifiers and connectives in terms of the primitive ones of L_0 (and L_1) have been given, and that "$(\forall x)Ax$" is a sentence of L_0. In what follows "$\Rightarrow$" stands for material implication and "$\vee$" for disjunction. As always in this book, and where needed, implicit quasi-quote conventions are in force.

I sometimes describe sentences of L_0 as canonical names even when they appear in sentential contexts. The square brackets flanking items within the scope of the predicates of Υ are *solely* devices of disambiguation, and aren't quotation operators.

Notice that in example (iv), "$(\forall x)Ax$," functions as a name: It appears in a nominal context. But in example (v) the very same sentence functions simply as a sentential component of another sentence. ("$(\forall x)Ax$" could play both roles—in different occurrences—in the same sentence of L_1.)

Because the same variable can appear in both nominal and sentential contexts, AU-quantification allows anaphoric linking between such contexts: This is the formal correspondent to the anaphorically unrestricted pronouns of 1.4.

L_1-Model Theory

Definition 4. A *model M for an L_1-language* has, as its domain Λ, an interpreted first-order language that's a (possibly proper) superset of L_0; M maps the constants of Υ to Λ, and the n-place predicates of Υ to the Cartesian n-product of Λ. "$=_{L_1}$" is interpreted as strict identity. M also maps the sentences of L_0 to themselves.

Three points. First: Allowing the domain of a model of L_1 to be an interpreted language properly containing the sentences of L_0 will contribute to the AU system's ability to handle endorsements of sentences from other languages. In what follows, I'll call these *language extensions of L_0*, or L_0-extensions.

Second: Despite allowing L_0-extensions as domains of models of L_1, the AU approach is predicative: The distinctive vocabulary of L_1 is *not* to appear in the domains of its models. What is to be said about impredicative approaches, specifically, the need for such, I say in Chapter 4.

Third: Moving from the context of L_0 to that of Λ isn't supposed to change the truth values of any of the sentences of L_0. One way to interpret L_0 is to (implicitly) supply it with a model M_0. If Λ is also interpreted by a model, Λ_M, then we can stipulate Λ_M and M_0 to have the same domains, and to agree on the vocabulary of L_0. It can be shown that the model theory thus restricted in models suffices for completeness.[5]

I now provide the model theory for L_1-languages in the familiar way to illustrate how close AU-quantification is to ordinary (objectual) quantification.

Definition 5. A *satisfaction mapping I* (with respect to a model M of an L_1-language) is a mapping from the set of L_1-variables to Λ; an *α-variant I'* (for L_1-variable α), of a satisfaction mapping I, is another satisfaction mapping which differs from I on at most $I(\alpha)$. For all satisfaction mappings I, $I(\tau) = M(\tau)$, if τ is either a canonical name or a constant of Υ.

---

[5] See Theorem 4 of the appendix of Azzouni 2001, 347.

Definition 6. A well-formed formula of an L_1-language is *satisfied by an interpretation I* iff one of the following six clauses applies to it:

(S1) If P is an n-place predicate of Υ, and $\tau_1, \ldots, \tau_n$, are L_1-terms, then $P[\tau_1] \ldots [\tau_n]$ is satisfied by I iff the n-tuple $\langle I(\tau_1), \ldots, I(\tau_n) \rangle$ is in $M(P)$.

(S2) If α is an L_1-variable, then α is satisfied by I iff $I(\alpha)$ is true in Λ.

(S3) If ϕ is a sentence of L_0 then it's satisfied by I iff it's true in Λ.

(S4) If ϕ is a wff of L_1, then $\neg\phi$ is satisfied by I iff ϕ isn't satisfied by I.

(S5) If ϕ and ψ are wffs of L_1, then $(\phi \,\&\, \psi)$ is satisfied by I iff ϕ and ψ are each satisfied by I.

(S6) If ϕ is a wff of L_1, and α an L_1-variable, then $(\Sigma\alpha)\phi$ is satisfied by I iff for some α-variant I' of I, ϕ is satisfied by I'.

These are, for the most part, pretty familiar. (S1), however, is slightly deceiving since, as already noted, among the L_1-terms (Definition 1) are the sentences of L_0. (S2) is unusual because it allows a freestanding variable to be satisfied or not. In its formulation (and in the formulation of (S3)), I have presupposed truth-in-Λ, but this isn't required: An alternative is to give models for both Λ and L_1, and then give the truth clauses for both languages simultaneously.

Just as Tarskian satisfaction drafts variables as temporary names of objects, so too, the AU approach allows a variable in a sentential context to act (temporarily) *sentence*like; just as variables, during Tarskian satisfaction, function as names for objects that the language itself may have no names for, so too do variables (in sentential contexts) stand stead for sentences *that may be inexpressible in L_1 itself.*

Blind Endorsements

The foregoing is enough to see how sentences of L_1 are interpreted, as well as how they express blind endorsements. Suppose there are sentences of L_0 on the blackboard in Room 201, and suppose Q (in Υ) is the predicate Written-on-the-blackboard-in-Room-201. Then "Every sentence on the blackboard in Room 201 is true" can be captured in L_1 this way:

$$(p)(Q[p] \Rightarrow p).$$

Given a singular term τ, we can express: the sentence τ refers to is true—like so:

$$(\Sigma p)([p] = [\tau] \,\&\, p).$$

In the case of Goldbach's conjecture, AU proof theory provides the following principle:[6]

$$(p)(q)([p] = [q] \to (p \Leftrightarrow q)).$$

---

[6] Via (P1)–(P8) in 3.3.

Since we know (contingently) that "Goldbach's conjecture" = "every even number is the sum of two primes," we have, using the above two sentences (where "Goldbach's conjecture" is substituted for "τ"):

> [Goldbach's conjecture] = [every even number is the sum of two primes] & every even number is the sum of two primes,

as desired.

I've just illustrated how L_1-languages handle blind endorsements without Convention T. Sticklers for tradition (*that* tradition, anyway) can have a closed statement that implies all the instances of the convention, if they want it. Axiomatize the truth predicate like so:

$$(p)(T[p] \Leftrightarrow p),$$

and then, by substitution of canonical names, all instances of T-biconditionals for the language L_0 follow. Notice, however, that the statement "$(p)(T[p] \Leftrightarrow p)$," says far more than that infinite list of T-biconditionals. For the resulting list is restricted only to the sentences of L_0, whereas "$(p)(T[p] \Leftrightarrow p)$" governs the sentences in any L_0-extension as well.

One thing we need to know is that L_1-model theory is adequate to supplying a theory of truth for *any* interpreted language L_0. It can't be that something in the structure of L_1-model theory rules out on its own an otherwise possible assignment of truth values to the sentences of L_0; equivalently, we want to know that the L_1-axioms of 3.3 are conservative with respect to the sentences of L_0: that we can't use them to provide additional implications of a set of sentences in the language of L_0 that can't be shown using the logic of L_0 alone. This needed result is provided by a corollary of Theorem 2 (from Azzouni 2001, 346). Indeed, that corollary gives us an even *stronger* result: Nothing in the structure of L_1-model theory rules out on its own an otherwise possible *model* for a set of L_0-consistent sentences.

3.3 Proof Theory for L_1-Languages

One advantage of the AU approach is that not only does it provide the logical form for blind endorsements but, more important, it's amenable to a complete proof system. I now give the axioms for the system of AU-quantifiers.

> Definition 7. A term τ is *substitutable* for free occurrences of an L_1-variable α in ϕ if (i) τ is a canonical name or an L_1-variable, or (ii) all occurrences of α are in nominal contexts; and (iii) if a variable, τ, isn't captured by a quantifier in ϕ.

> Definition 8. Let $\alpha_1, \ldots, \alpha_n$ be any n L_1-variables. Then $(\alpha_1) \ldots (\alpha_n)\phi$ is a *generalization* of the wff ϕ.

> Definition 9. Where α and β are L_1-variables, and ϕ and ψ are wffs, the L_1-*axioms* are all wffs (and generalizations of such) of the following forms:

> (P1) All tautologies.

(P2) All logical truths of L_0 without free variables.[7]

(P3) $(\alpha)\phi \Rightarrow \phi^\tau$, where ϕ^τ is gotten from ϕ by the substitution of a substitutable term τ for all occurrences of α in ϕ.

(P4) $(\alpha)(\phi \Rightarrow \psi) \Rightarrow ((\alpha)\phi \Rightarrow (\alpha)\psi)$.

(P5) $\phi \Rightarrow (\alpha)\phi$, where α does not occur free in ϕ.

(P6) $[\alpha] = [\alpha]$.

(P7) $[\alpha] = [\beta] \Rightarrow (\phi \Rightarrow \phi')$, where ϕ is atomic, and ϕ' is obtained from ϕ by replacing α in zero or more (but not necessarily all) places by β.

(P8) $\neg([\phi] = [\psi])$, where ϕ and ψ are distinct sentences of L_0.

Definition 10. The inference rule is modus ponens, and the notion of a deduction is the standard one.

Notice how similar the proof theory is to that of ordinary first-order logic. This makes it easy to understand, despite (i) the appearance of variables in both nominal and sentential positions, and (ii) the corresponding function of the sentences of L_0 both as sentences and as names for themselves in L_1. Only (P8) is unusual; but its import is transparent: If two sentences differ then, of course, their names don't refer to the same things.

Examples. Here are two L_1-deductions.[8]

　　1. $(p)(q)(p \Rightarrow (q \Rightarrow p))$, P1

and the second is:

　　1. $(p)(T[p] \Leftrightarrow p) \Rightarrow (T[(\forall x)Ax \Rightarrow (\forall x)Ax] \Leftrightarrow ((\forall x)Ax \Rightarrow (\forall x)Ax))$, P3

　　2. $(\forall x)Ax \Rightarrow (\forall x)Ax$, P2

　　3. $1 \Rightarrow (2 \Rightarrow ((p)'T[p] \Leftrightarrow p) \Rightarrow T[(\forall x)Ax \Rightarrow (\forall x)Ax])]$, P1

　　4. $2 \Rightarrow ((p)((T[p] \Leftrightarrow p) \Rightarrow T[(\forall x)Ax \Rightarrow (\forall x)Ax)])$, 1, 2, mp

　　5. $(p)(T[p] \Leftrightarrow p) \Rightarrow (T[(\forall x)Ax \Rightarrow (\forall x)Ax])$. 2, 3, mp

Another point should be stressed about the L_1-axioms. The system is first-order in a respectable sense: The completeness proof given for the system in my 2001 shows that the logic doesn't involve a noneffective notion of derivation, and, related to this, that nothing infinitary is implicitly involved in the notion of validity for AU-quantifiers.

One might have thought otherwise, that the presence of language extensions shows that what's going on is similar to what goes on when substitutional quantification is supplemented with language extensions to regain the completeness proof.[9] But that's not the appropriate analogy. Rather, the analogy

---

[7] Since the logical truths of L_0 are recursively enumerable, the resulting set of axioms is recursive despite this clause.

[8] As in 3.2, "$(\forall x)Ax$" is taken to be a sentence of L_0.

[9] See Dunn and Belnap 1968. Also Leblanc 1983.

should be drawn to ordinary "objectual" quantification, where a statement like "$(x)Px$" may be true without there being a substitution instance "Pb" that's true as well. What may fool us into thinking in terms of the substitutional-semantics case is (i) the long tradition of handling quantification into sentential position substitutionally, plus in this particular case, (ii) the presence of the sentences of L_0 in every model. But as the presentation of pure L_1-languages indicates (in 3.4), L_0, along with the canonical names for its sentences, can be easily excised from the system. It is, therefore, an example of pure quantification into sentential *and* nominal position; and completeness is a natural result of that fact.

Truth Conditions

Since the L_1-axioms license blind-endorsement inferences only via inference rules governing AU-quantifiers, this provides a crisp distinction between the logic of blind endorsement and substantial semantic principles that go beyond this logic. Furthermore, if we choose to provide substantial semantic principles, this can be easily done in a way that mimics Tarski-style truth conditions for a theory. Here's an illustration.

Let L_0 be any first-order language interpreted in any model. Let L_1 contain L_0, and let Υ contain the predicates $C[p][q][r]$ (r is the conjunction of p and q), $N[p][q]$ (q is the negation of p), $E[p]$ (p has an initial existential prefix), and $V[p][q]$ (q is gotten from p by the substitution of a name for all occurrences of the free variable α in the well-formed formula gotten from p by the stripping of its initial existential prefix $(\Sigma\alpha)$). Here are truth conditions for the sentences of L_0.

(SM1) $(p)(q)(r)\ (C[p][q][r] \Rightarrow (r \Leftrightarrow (p\ \&\ q)))$.

(SM2) $(p)(q)\ (N[p][q] \Rightarrow (p \Leftrightarrow \neg q))$.

(SM3) $(p)(E[p] \Rightarrow (p \Leftrightarrow (\Sigma q)(V[p][q]\ \&\ q))$.

These are shockingly simple.[10] Notice the absence of base clauses for sentences without quantifiers or connectives.[11] They aren't needed: SM1–3, as they stand, "chase truth up the tree of grammar," as Quine has so nicely put the requirement on semantics; they explain how the truth values of conjunctions, negations, and existentially quantified sentences depend on the

---

[10] In part this is because axiomatic constraints on the syntax of L_0 have been omitted. Concatenation theory or a fuller set of predicates and axioms secures the syntactic properties needed for metalogical investigations (e.g., definitions of *formula*, *closed formula*, the uniqueness of the conjunction of two formulas, etc.). The point of SM1–3 is to illustrate what the "truth" clauses look like in an L_1-language.

[11] The base clauses of the Tarskian approach have generated the most controversy in the literature, as I indicated in 1.7. They can be omitted on the AU approach because they don't rely on a use of blind truth-endorsement but only on redundant T-biconditionals or something similar. On an approach where redundant uses of a truth predicate are replaced by the sentences themselves, the mere presence of the sentences of L_0 in L_1 suffices for semantic purposes—subject to the discussion in 1.7.

truth values of other sentences. If the model interpreting L_0 is one in which every item in the domain has a name, SM1–3 recursively define the truth of the sentences of L_0 in terms of other sentences of L_0. Otherwise, the needed names are in L_0-extensions. Incidentally, nothing requires (all) the constants needed to satisfy SM1–3 to belong to one (particular) L_0-extension: Regardless of the cardinality of the model interpreting L_0, no L_0-extension need have more than a finite number of additional constants in order to satisfy SM1–3.[12]

Two final points about the proof theory. Much has been made of the question of whether a (deflationist) theory of truth should be conservative. To some extent, the debate is ill defined. The deflationist—or one version of such, anyway—claims that truth doesn't pick out a substantial property, and it's been suggested that a necessary condition on insubstantiality-as-such is that the deflationist truth predicate be conservative in the sense that no additional theorems in the language of a theory be derivable by means of the theory of truth + that theory, beyond what's derivable from the theory alone.

The claim has *some* plausibility given a commitment to redundancy views of the truth predicate. But on a view where the predicate is a device for navigating use and mention, it is *not* obvious why such a thing should prove conservative. In any case, standard illustrations of how the truth predicate fails to be conservative in no sense involve just the truth predicate.[13] So it should be no surprise that the conservativeness of AU-quantifiers, when added to any first-order language, is easily established.[14] This can change—and *should* change—when substantial principles governing the predicates and constants of Υ are added to AU-quantifier theory.

This brings us directly to the second point. AU-quantifier theory is complete and, in strict analogy to first-order logic, all bets are off regarding completeness, if intended interpretations are given for nonlogical predicates and constants.[15] The lesson to be drawn is that the syntactic and semantic theory—of specific languages or classes of such—should be treated along the lines of first-order theories of specific predicates and constants with intended interpretations. Truth—insofar as its blind-endorsement role is concerned—should be assimilated to AU-quantifiers so that it's a matter of logic *proper*. It's worth repeating: The neat togetherness of syntax and truth in Tarski's classical presentation has led to the tainting of the concept of truth with elements strictly foreign to it.

---

[12] The foregoing also illustrates something I've (up to now) suppressed explicit mention of in this chapter: Describing AU-quantifiers Tarski-style isn't a requirement on their presentation. For their truth conditions—being truth conditions—can be given *solely* by clauses in terms of AU-quantifiers themselves, connectives, and some mathematical apparatus. In this sense, AU-quantifiers are "self-interpreting": they do *not* require a characterization in terms of Tarski-truth.

[13] See Shapiro 1998 and my (1999) response. Or see 9.9.

[14] See the corollaries to Theorem 2 of Azzouni 2001, 346–47.

[15] That is, first-order theories can fail, or not, to be complete, e.g., PA or Presburger arithmetic.

3.4 Pure L_1-Languages

L_1-languages facilitate endorsements of the sentences in foreign languages via (i) language extensions, coupled with (ii) quantification into sentential position, together which allow endorsement of foreign sentences without such sentences appearing in the language L_0 or even in the language L_1, that is, the language of the endorser. Just as the truth of "$(\exists x)Px$" is compatible with there being no constant b available so that Pb is true (in the standard first-order case), so here we find that a commitment to $(\Sigma p) \ldots p \ldots$ is compatible with there being no sentence S the speaker has access to (that can be substituted for "p") so that $\ldots S \ldots$ is similarly committable to. Thus, the tethering (by immanent truth theories) via Convention T (the explicit list of T-biconditionals) of blind endorsement to the capacity—at least in principle—to directly present the sentences, or translations thereof, so committed to, is gone: The anaphorically unrestricted quantifier extends genuine quantification from "mention" to "use."

I've also indicated, although briefly—by the giving of SM1–3—how truth conditions can be supplied when these are to be linked to endorsements. It can prove valuable to supply such truth conditions for languages that, although similar to ours in their logic, aren't so similar in their grammar. We may want, that is, to characterize the "truth conditions" of such a language in terms of its syntax; here "truth conditions" are understood in terms of the AU-quantifiers. But L_1-languages contain intended L_0-languages in the sense that every model of a particular L_1-language contains the sentences of a specific L_0-language, and canonical names of the L_1-language name the same sentences of L_0 in every one of its models. Characterization of the syntax of a foreign language will be easier if clauses describing the syntax of the sentences of L_0 aren't included. The following characterization of a pure L_1-language is meant to facilitate this.

> Definition 11. A *pure* L_1-language is an L_1-language without canonical names. The other definitions (of *pure model, interpretation*, etc.) are the same for pure L_1-languages except for the removal of canonical-name clauses, and the expansion of pure models to include arbitrary first-order languages.

Notice that pure L_1-languages are still such languages with respect to a given L_0; it's just that the sentences of L_0 don't appear in every domain of a model for an L_1-language. Thus, as Definition 12 makes clear, (P2) still holds in the proof theory of such languages.

> Definition 12. The pure L_1-axioms are the same as the L_1-axioms except that (P8) is replaced with:
>
> (P8*i) $(\Sigma \alpha)\alpha$,
>
> (P8*ii) $(\alpha_1) \ldots (\alpha_n)(\Sigma \alpha)(\neg([\alpha_1] = [\alpha])$ &$\ldots$& $\neg([\alpha_n] = [\alpha])$ & $\alpha)$,

(P8^i) $(\Sigma\alpha)\neg\ \alpha,$

(P8^ii) $(\alpha_1)\ldots(\alpha_n)(\Sigma\alpha)(\neg([\alpha_1] = [\alpha])\ \&\ldots\&\ \neg([\alpha_n] = [\alpha])\ \&\ \neg\alpha).$[16]

Two last points about pure L_1-languages: First, (P8*) and (P8^) constrain possible models of a pure L_1-language to contain countably many distinct true sentences and false sentences. No *further* conditions on the syntax or semantics of such languages are given. If wanted, these are to be additional axioms on the nonlogical vocabulary. Second, pure AU-quantifier theory doesn't admit *any* T-biconditionals since canonical names are absent.

3.5 Sentences with Demonstratives

It's important to recognize the great (semantic) divide between substitutional approaches and the version of genuine quantification on offer in the AU approach. One way to illustrate this divide is to consider a kind of sentence that has given substitutional approaches to truth a headache: sentences with demonstratives.[17]

Let's approach this informally. Suppose John says, "I am tall," and suppose I say, "What John says is true." My remark goes into Anaphorish as, "There is something such that *it* is what John says and *it*." But (here's an objection) whatever *it* is, *it* can't be "I am tall," since it's not the case that there is something such that *it* is what John says and I am tall. (Suppose I'm $5'2''$.)

This objection turns on seeing the quantifier "There is something . . ." as having truth conditions that operate by implicit *substitutions* of (tokens of) sentences for the pronoun "it." That is, John has uttered a token of a sentence, and on substitutional approaches, another token of whatever it is that John has expressed must successfully replace "it" in order for my remark to have the right truth value. The problem is that substituting something syntactically identical to what John said won't work.

Ambiguity calls for a finer division among tokens than shape.[18] One might think this drives us to bring content into the mix: Sentence-tokens are equivalent, if they "mean" the same thing—express the same *proposition*, say. Now, despite the discussion in 2.9, I'm confident that one needn't rush to an invocation of content—at least not on the basis of this concern. A finer division among tokens, having to do with their relations to (say) utterers, or more broadly, to place and time, will suffice. That is to say, we can track the needed

---

[16] My thanks to Elliot Mendelson for pointing out the need to include generalization-free versions of these axioms.

[17] My thanks to Douglas Patterson for pressing issues that led to my writing this section, and to Mark Richard (1996) for his discussion that kept worrying me.

[18] Actually, the matter is more subtle: verbal and written versions of the same (standing) sentence hardly have the same "shape." Still one presumes (apart from issues of ambiguity) that the individuation conditions on sentences—when two tokens are tokens of the same sentence—can nevertheless be given in physical terms.

divisions among the tokens of the same syntactic types not directly in terms of content or meaning, but in terms of (broadly speaking) relational physical facts: By invoking a bit more of the history of the tokens, we will be able to manage the needed finer division among tokens of the same syntactic types.

But demonstratives pose a more subtle problem because what's called for is more than a finer division among tokens of the same syntactic objects. What's needed to make good on the truth conditions for "What John says is true," (as uttered by me) is a token that isn't—even broadly construed—of the same physical shape as what John said. For John used the word "I," and I must use "he," or something else. Furthermore, the tokens, "I am tall," as uttered by John, and (say) "he is tall," as uttered by me, are not, generally, intersubstitutable in the same contexts. Rather, it looks as if some story must be told about how the statement uttered by me tracks the content—at least when uttered by me—of the statement uttered by John.

Now, of course, it's open to the proponent of a substitutionalist approach to truth to assume that all this is managed *somehow* without taking himself to be required to give the details—perhaps by claiming that actually spelling all this out isn't a required part of the theory of truth. Depending on what other sorts of doctrines the substitutionalist wants to hold, this is something that could cause trouble later. For example, one may fear that, in the final analysis, he has to offer a story that relies on such utterances tracking each other's truth conditions; and that the story of truth conditions required has to be a substantial one.[19]

In any case, there is still a big problem because it looks as if the substitutionalist has to supply at least one other token that can function in an utterance of *mine* so that "What John said is true," as uttered by me, will have the right truth value. And so it seems that the situation is much worse than with the earlier case of ambiguity: A finer division among the tokens of sentences is something one can see can be done, and probably done without directly invoking content. But this subtle tracking to be done by utterances of very different syntactic form is a different matter: One can argue that we have no reason to believe the substitutionalist's truth-theory offering can work for demonstrative-laden sentences unless we are told enough to see that the needed tokens exist.

Just this last bit is what the genuine quantification on offer through the AU approach *doesn't* require. Returning to the original example, recall, John says, "I am tall." I say, "What John says is true," or "There is something such that *it* is what John says and *it*." What *it* is that? The sentence (-on-a-sense, perhaps), of course, *a token of which was uttered by John*. AU-quantification—to give the right answers for the truth values of its blind endorsements—doesn't require that any

---

[19] Caveats. This needn't trouble the BTDist of chapter 1.7, as I argued there. But it may be troublesome to the MTDist if, for example, such tracking needs to be done in terms of *objects* that demonstrative terms must directly refer to. I'm optimistic that the work needed doesn't require objects, if only because such apparatus is fully present and functional in fictional discourse, where (so I claim) no such objects are to be found. But all of this, of course, is quite controversial.

token (tracking the content of John's utterance) successfully *substitute* for "it" in my utterance; more remarkably, it doesn't even require that the statement expressed by John's utterance have more than one token. What it needs is only that the variable in sentential position successfully carry the "force" of assenting to the utterance the variable refers to. Only if one is overly focused on T-biconditionals, which give the impression that one can only either assert that a sentence is true or *use* that very sentence, will it be thought that a variable can stand stead for an utterance only by—as it were—presenting the content of John's utterance in a way that requires substitution in a very different sentence with the same "content" (e.g., by substituting that very utterance for the variable, and adjusting the demonstratives accordingly).

The analogous suggestion, with ordinary objectual quantification, is the thought that an ordinary variable, when standing stead for an unnamed object must—somehow—duplicate that object's properties. No, in both cases: All that's needed is that the semantic values assigned to sentences containing that variable be linked in the appropriate way to what that variable is standing stead *for*. It's worth adding that in the case of ordinary objectual quantification, bound variables themselves don't carry "content"—unless "content" is understood as sheer reference to something. All the work is done by the context the variable appears in. For example, "$(\exists x)(Rx \ \& \ Qx)$" singles out something (if it does) by virtue of "R" and "Q."[20]

The foregoing reveals an interesting advantage that AU-quantification has over substitutional approaches. It's a bit odd to imagine this, perhaps, but suppose that John utters something which expresses a statement that can be expressed by no other person under any other circumstances. The substitutional approach, requiring a token to track that statement—as uttered by others—can't assign blind endorsements on the part of others (to John's statement) the correct truth values; but this is no problem for the AU approach. Similarly, imagine you believe that demonstrative-laden utterances express sentences that can't be expressed by demonstrative-free utterances (people *have* believed this). Or, more drastically, imagine you believe that (certain) demonstrative-laden utterances express propositions uniquely: No other utterances can mean the same thing. In such cases, only the AU approach—which doesn't require tokening of sentences to fix the truth values of blind endorsements—can underwrite blind endorsements to such things.

This puts the proponent of AU-quantification on stronger ground when she resists telling a story about content that explains how sentences of very different syntactic forms can stand stead for each other in our utterances. But is the AU approach compatible with something as strong as the Quinean hostility to a coherent notion of content? I think it is (although, again, the

---

[20] Note the contrast with (some) pronouns in ordinary English. These days "he," even when bound, is taken by many people to carry the content of male gender. By contrast, AU-variables are like the variables of ordinary objectual quantification: A freestanding variable "*x*" in sentential position sententializes *whatever* (if anything) is picked out by the contexts of all the variables it's cross-linked to.

approach hardly *requires* meaning-scepticism). How finely we should individuate equivalence classes of sentence tokens is a matter that turns on usage,
and when speakers resist "same-saying" construals of what they claim. One
needn't think there is anything systematic going on here. And, for purposes of
AU-quantification, every utterance with demonstratives can be treated as
a statement with a token of only one. Of course, for broader purposes of handling ascriptions of what's said to one another by *paraphrase* (e.g., John says:
"I'm hungry," Jack says, "John said he's hungry," or "He said he's hungry"),
we may, in the final analysis, need to invoke a richer notion of meaning, and for
that we may need truth conditions—equally richly construed—to anchor the
semantics of such ascriptions. But maybe not.

3.6   *Concluding Remarks*

L_1-languages show how to blindly endorse sentences from foreign languages
without either presuming (falsely) that such sentences are translatable to ones
of ours, or presuming (falsely) that we've a grip on propositions the expression
of which transcends the resources of our own language. As long as the context
into which AU-quantifiers quantify requires nothing more of foreign sentences
other than their being truth vehicles, that such sentences are foreign to L_1
proves benign. If the contexts require more than truth or falsity, the AU
approach needs supplementation.

Modeling the transcendent truth predicate, as it occurs in ordinary language, via AU-quantifiers, requires breaking ranks with biconditional truth
deflationists. Although we can both agree on Quine's original claim that the
truth predicate is only for semantic descent and ascent, the proponent of AU-
quantification must—ultimately—deny the centrality of the T-biconditionals.
That a truth predicate can be introduced, once AU-quantifiers are in place
with the simple formulation: $(p)(T[p] \Leftrightarrow p)$, only provides a faint echo of the
claim that the truth predicate is governed by T-biconditionals, since not only
does that statement outstrip the list of T-biconditionals in what it expresses,
as I said earlier, but it also expresses something that seems, strictly speaking,
grammatically beyond the resources of ordinary English altogether.

But this raises an issue. Given that AU-quantifiers show that what a
transcendent truth predicate needs to do can be done, one can ask: If AU-
quantifiers themselves aren't in natural languages (and they seem not to be),
how is the trick done in natural languages to begin with? We seem, in those
languages, to be stuck with a predicate, and with nominal quantification over
propositions; and I've argued that those can't provide blind endorsements of
foreign sentences. So is the apparent transcendence of the truth predicate—in
English—a grammatical illusion? I'll discuss this in the conclusion to part I.

Apart from this concern, some philosophers also worry that unless a
logical device can be translated into the mother tongue, in this case ordinary
English, it can't be understood. I've denied this. I've denied that formulations
in L_1 are even *hard* to understand—let alone incomprehensible; and I've

claimed the same is true of the anaphorically unrestricted pronouns of 1.4. There's nothing difficult in understanding how such devices warp anaphora, provided (of course) one grasps anaphora to begin with. This provides a clue to how regimentations of ordinary languages are both valuable and result in modifications that we can nevertheless still understand the meaning of. I say more about this in chapter 4.

Related to this is another issue that I've so far suppressed discussion of. This is the so-called liar paradoxes—a topic Tarski (1983b) took up right at the beginning of his seminal paper. It turns out that the status of such paradoxes in natural languages is related to the question of what regimentations of natural languages are needed for. These are the topics of the next two chapters.

4

Regimentation
and Paradox

4.1 Motivations for Regimentation

In 1.3, I noted the difficulty natural languages have with anaphora. Thus AU-quantifiers couldn't have emerged spontaneously in natural languages because of reasons *extrinsic* to blind-endorsement needs: the confusion such quantifiers would cause. If neat devices (e.g., variables or schematic letters) are missing, anaphora must be kept to a minimum.[1]

Armour-Garb and Beall (2001, 601) cleverly apply Gould and Lewontin's (1978) notion of a spandrel to liar paradoxes, describing them as "semantic spandrels"; they write that liar paradoxes are sentences "which arise as by-products of introducing 'true' into a language with the underlying grammatical rules that exist in English." Despite the Lamarckian evolution of natural languages, there are *many* (traditional) evolutionary lessons applicable to those languages. One is the engineering point that evolutionary structures are jerry-built on structures already in place for a sequence of quite different (and historically successive) reasons.[2] Thus linguistic devices from the vernacular often

---

[1] Uniform convergence is very difficult for novice calculus students because of the quantifier interchange involved, and *that's* difficult because natural languages don't equip ordinary speakers to handle complicated cases of anaphora.

[2] Larmarckian evolution allows (at least in principle) the ditching of previous structures altogether, and starting anew. But there is great inertia in the evolution of ordinary languages if only because of constraints imposed by the neurophysiological constants of the human brain.

have properties that are, given their current functional roles, irrelevant.[3] And, despite the importance of the empirical study of the syntax and semantics of natural languages, this is why *regimentation* (in something like the Quinean sense) is still of interest.

A regimentation, as I understand it, of a designated section of ordinary language *replaces* that designated section with a (piece of) engineered artificial language, not in the sense of giving speakers a different language to (as a practical matter) use, or speak in, but more narrowly, of giving *normative constraints* on inferences, and other logical matters, that speakers *should* acknowledge on the basis of statements they've committed themselves to.[4] The model for regimentation, that is, is very close to how highly developed branches of mathematics are used to correct the nonprofessional's computations (in arithmetic, say, or in probability). What's claimed isn't (and can't be) that the ordinary person *really means* to carry out calculations of such and such a sort—and that, as an empirical matter, the rigor introduced into the development of a mathematical field makes the ordinary person's intentions explicit. No such empirical study (of the ordinary person's "intentions") is carried out by mathematicians; and when psychologists do study how the ordinary person computes, what's found instead is that the dispositions of the ordinary person usually deviate sharply from the mathematized approach to a subject area.[5] Rather, rigorous mathematics constrains more informal mathematical practice by serving as a corrective when disputes arise.

Regimentation, as philosophers and logicians practice it, is (as I've said) similar. The purpose is (i) to systematically present sentential vehicles with computationally transparent and tractable inferential properties, and to supply a mathematically tractable semantic theory for such items. (E.g., first-order languages have mechanically recognizable derivation rules, and the semantics for such languages are well understood.[6]); and (ii) to use results about such systematically presented sentential vehicles—*if they are attractive and clear enough*—as the final court of appeal regarding logical issues about the ordinary-language statements the regimentation concerns. The italicized rider is crucial: Regimentations come and go: They don't constrain decisions, say, about what logically follows from what, if their dictates are particularly

---

[3] The analogy with artificial design versus evolved design is exact. Despite the subtlety of evolved structures, there is always the likelihood that aspects of it are useless or even pernicious (given its current niche), and that something else, designed from scratch, could do the job much *better*. An example is (alas) our *brains*—with respect to number crunching, but not only that, I'm sorry to say.

[4] Don't overrate my invocation of normativity. I say more shortly, but I need to stress right away that what's "normative" about regimentation is only that we use it to *stipulate*, for example, that "so-and-so" (from the vernacular) means "***," where what "so-and-so" means in the vernacular is either unclear or clear enough—but we find it more valuable to mean something somewhat *different*. "Stipulate," I also want to stress, is the right word to describe in what sense a regimentation trumps the dictates of the vernacular it regiments.

[5] See Dehaene 1997 for some examples of this.

[6] Consistent models for such languages aren't, generally, mechanically recognizable; but there are mathematical results about such models that give us a fairly good grip on their properties.

controversial—for example, if important functional roles that idioms in ordinary language implement are ones the regimented proxies of those idioms are incapable of. Thus, the use in regimentations of, for example, first-order classical languages, is (i) for technical reasons—our superior facility with them (drawing inferences within them and proving properties of them)—and (ii) because they adequately replicate the functions and roles of idioms in ordinary language that we want to keep.

It's easy to confuse regimentation with an empirical study of natural languages. Under such a misapprehension, one presumes that if a regimented idiom operates the way an idiom in a natural language does, then the syntax and semantics of the regimented idiom may be (provisionally) attributed to the natural-language idiom. This is naïve: The idioms of invented languages can suggest empirical hypotheses about empirical idioms that we can then try to vindicate the presence of. But the purpose of regimentation is different: It is, as I've said, a *normative practice*. (i) The regimenter discovers the needs that an ordinary language idiom **O** satisfies in ordinary language. Success in fulfilling these needs may turn only on a subset of the properties of **O**. Or it may be that—from a technical point of view—the best way to implement those needs is with a device with none of **O**'s properties. It can also turn out that **O** has two or more roles, and that these are—again from an engineering point of view— best segregated, so that some of the uses **O** has are ignored (given certain purposes). (ii) The regimenter coins logical devices with mathematically transparent semantic properties to satisfy those needs. "Mathematically transparent" means that the properties of such designed devices are matters of pure mathematics: There is no nonmathematical issue of what properties they "truly" have. (iii) (And it's here the normative element in regimentation is explicit:) The logical properties of such coined devices *trump* the empirically determined properties of the original items in natural languages *in some cases* and for some purposes. For example, certain apparently semantic intuitions of ordinary speakers may be ignored—e.g., certain implications that speakers ordinarily presume are disallowed; or certain implications speakers ordinarily don't accept are taken nevertheless to be implications of (some of) the sentences (such idioms appear in).[7]

One important condition on regimentations is that they not lead to practical impossibilities. As I've already indicated, natural languages have species-specific syntactic structures that make them easy to learn and use by members of our species. These species-specific syntactic structures aren't present in the syntax of logical systems, if only because the implicational properties of sentences in

---

[7] Some philosophers of language attempt to explain away certain linguistic intuitions— describing them, for example, as "performance errors" or "pragmatics" and not "semantics." This *may* be appropriate if one really is designing a theory about the idioms of natural languages (which—given the formal tools nearly all philosophers of language automatically help themselves to—*can't be* how to construe what they're up to); but the regimenter's aims are different: The regimenter recognizes that natural-language idioms can involve all sorts of linguistic practices that give rise to intuitions that are, strictly speaking, irrelevant to the functions that the idioms in question have been discovered to have.

such logical systems (when complete) are reflected proof-theoretically. Ordinary languages are quite different, but even apart from this difference, it's likely—as I've already indicated—that there are strange jerry-built aspects to the syntax of ordinary language that reflect not the design of a linguistic engineer of genius—Frege, say—but the neurophysiological needs of the users of that language that, in turn, are the weird contingent historical products of evolution.

Given, therefore, our inability to desert the vernacular without serious neurophysiological mutation, what should a regimentation look like? Consider (again) mathematics as an excellent example of how regimentation *already* works. The practice of mathematics occurs in ordinary language—as it must—but with two significant differences. First, the vocabulary of ordinary language is routinely supplemented with additional terms indispensable for the mathematical tasks at hand. But, more important, *deviations* from ordinary language are licensed. These deviations must be practically implementable—in the sense that someone who uses ordinary language (and continues to use ordinary language) will be able to learn to correct ordinary language in the respects that the regimentation requires. It's in this sense that a regimentation functions regulatively: If we introduce a regimentation that brands certain inferences in ordinary language as "wrong," but as a practical matter we can't correct our ordinary-language practices to neatly remove those inferences, then we're barred from regimenting such "infelicities" away.[8]

4.2 The Horseshoe as an Example of a Regimentation

Consider the "horseshoe," "$\Rightarrow$," of classical logic and its truth table. There has been much discussion over the years about the misfit between the ordinary English "if—then," which isn't truth-functional, and the horseshoe, which is. One way the horseshoe fails to fit ordinary English is that statements of the form $(\mathbf{A} \Rightarrow \mathbf{B})$ are true when the antecedent $\mathbf{A}$ is false. This, in turn, allows generalizations of the form $(x)(\mathbf{P}x \Rightarrow \mathbf{Q}x)$ to be true when there are items that $\mathbf{P}$ doesn't hold of (although $\mathbf{Q}$ holds of anything $\mathbf{P}$ *does* hold of). Philosophers sometimes justify regimenting "if—then," with the horseshoe by the purported convenience and simplicity of truth-functionality. But this, actually, is pretty dubious, if only because the decision-procedural advantages of truth tables, in practice, don't buy very much: The decision procedure in question is very time consuming and of limited applicability; this also holds of strategic time-saving modifications of it. The real advantages of this regimentation lie elsewhere, as I show.

Consider a simple language with sentential variables, $\mathbf{A}$, $\mathbf{B}$, $\mathbf{C}$,..., parentheses, and the single two-place connective, &; imagine that its sentences are the sentential variables and any sentence of the form $(S_1 \;\&\; S_2)$, where S_1 and S_2 are themselves sentences. As those familiar with such matters say, it's

---

[8] This aspect of the normativity of regimentation is shared with ethics: ought implies can.

an easy induction to show that every sentence of this simple language has an even number of parentheses. One variant of the proof goes this way: The base step is to consider sentences of length one; but these haven't got parentheses at all, and so the even-parenthesis thesis *holds vacuously* of sentences of length one.[9] One then takes the induction step. That a theorem can hold vacuously of some of the cases it applies to turns precisely on assimilating the "if—then" of the vernacular to the horseshoe.

Two points. First, and this is why I give this example at all, the budding mathematician, although taken aback the first couple of times he or she learns that vacuous cases of this sort can be used as steps in a proof to establish a theorem,[10] will eventually learn to override the usages of "if—then" that don't allow vacuous cases to be subsumed this way (he or she learns, that is, to override the non-truth-functionality of "if—then," in these cases). This is, as a practical matter, easy to do, and so the truth-functional regimentation normatively trumps ordinary language, if there is an advantage in allowing it to so trump.

But what *is* the advantage, exactly? I've already suggested it isn't that truth-functional connectives are, generally, easier to reason with than non-truth-functional connectives: One has to *learn* the trick of subsuming the vacuous cases under the general case in the way that the truth functionality of the horseshoe allows; it *doesn't* come naturally. But it's precisely the subsumption of vacuous cases into a neat generalization that's the point of doing this in the first place—the number of *easily stated* generalizations are increased as a result: One no longer has to state the (vacuous) exceptions explicitly: The horseshoe does it for you![11]

4.3 *Regimenting Truth: Preliminaries*

Before focusing specifically on what's called for in a regimentation of the truth idiom, some jargon is needed: When an idiom (or sentence) is regimented, I'll describe the result as a *proxy* of the original idiom or sentence in natural language.

Turning to the vernacular term "true," I've already claimed that the role of that idiom in the vernacular is part of a package deal the function of which, and the only function of which (that we need concern ourselves with), is to

---

[9] Another way to go is to say that 0 is an even number; but that trick is usually unavailable to reclassify vacuous cases.

[10] Vacuous cases arise anywhere in inductions, not just at the base step: One may divide the inductive step into several kinds of cases, and in some of these the theorem may hold vacuously.

[11] Easily stated generalizations are hard to find; *anything* that increases our ability to express them thus increases our grasp of a subject matter—it increases what we can say, and what we can (explicitly) prove about that subject matter; more generally, it increases what we can see clearly about a subject matter. In this case the increase in generalizations neatly expressible is achieved by what looks like a cheap notational trick. But one should *never* underestimate the power of cheap notational tricks, for example, letting "0" function as a placeholder.

handle blind endorsement.[12] I've also already suggested that the introduction of a *truth predicate* in ordinary language for these purposes is an accident of linguistic history: Anaphora isn't handled in ordinary language by *variables*, but rather by a cacophony of heterogeneous devices that make it forbiddingly complicated to introduce links between sentential and nominal positions. From this fact, however, yet another spandrel emerges—one that has long bedeviled philosophers: the (intuitive) centrality of the T-biconditionals. AU-quantifiers, as we've seen, don't need T-biconditionals to facilitate blind endorsements because an AU-quantifier can simultaneously bind a freestanding variable, and one occurring within a complex description. Automatically, what we endorse (in uttering, say, "$(p)(Q[p] \Rightarrow p)$") is anything that falls under "Q." This blind-endorsement task is straightforwardly implementable by the AU-quantifier. But, in the vernacular, a truth predicate, being a predicate, must be axiomatically fixed in what falls under it (otherwise what falls under "Q" needn't fall under "T" in "$(\forall x)(Qx \Rightarrow Tx)$," even if it *should*). In itself, that's no big deal. Unfortunately, because what falls under such a predicate cannot be given by a description,[13] it must be done on a sentence-by-sentence basis: Thus the unnatural listlike quality of the set of T-biconditionals—when unaccompanied, I mean, by a theory they can be deduced from.[14]

So far, I've applied "spandrel" to the presence of a truth *predicate* in ordinary language; I've also, as a result, applied "spandrel" to the intuitive centrality of T-biconditionals to our notion of truth. I now want to claim that one additional set of spandrels exists in ordinary language. These are sentences which needn't appear *at all*[15] in our regimentations of the ordinary language. Before giving specific examples of such sentences, I should first make a methodological point and introduce some terminology. When an idiom is regimented, it can be that its scope is narrowed in such a way that (certain) sentences within which that idiom appears (in the vernacular) aren't replicable in the regimented language: They have no proxies. Such sentences have an interesting twilight status: Since they are present in the vernacular, they—let's say—may be uttered by speakers on certain occasions. But (and here the normativity of regimentation again manifests itself), we nevertheless regard such sentences as failures of a sort; we don't regard speakers as having successfully uttered something. We can say, in fact, that speakers have failed to express a *proposition*.

---

[12] See chapter 2. Of course, there may also be other (e.g., rhetorical) purposes that "true" is put to that I take the regimenter as free to ignore.

[13] Recall the discussion in 1.8 for indications for why the MTDist is right.

[14] By contrast, chapter 3 showed that to axiomatize AU-quantifiers, one can (pretty much) imitate the axiomatizations of ordinary quantifiers.

[15] The intuitive *centrality* of T-biconditionals, so I claim, is a spandrel due to the role satisfied by AU-quantifiers in our regimentation being one satisfied in natural languages by a truth predicate that *needs* to be fixed by such biconditionals to what it applies to. But I'm not claiming that T-biconditionals *themselves*—the sentences, I mean—are an (eliminable) spandrel. For, as we saw in chapter 3, a predicate satisfying those biconditionals is certainly definable in AU-languages (under certain conditions). Not so for the set of sentences I go on to discuss next.

This use of "proposition" is purely technical. Denying that a sentence of the vernacular "expresses a proposition" in this sense doesn't mean that the sentence is meaningless *in the vernacular*. Presumably, anything successfully expressed in the vernacular isn't meaningless; and indeed, there still remains the empirical question to be answered about the semantic (and syntactic) properties of what's been expressed in the vernacular. But if it has been determined that, given the roles of the idioms appearing in that sentence, resources allowing the construction of its proxy needn't be included in the regimentation, we see the question about the semantic and syntactic properties of the sentence (in the vernacular) as only an empirical question *about the vernacular*; we don't see it as a question that *we* have to answer in order to determine (say) what the implications of what we should have (normatively speaking) claimed, or not claimed, about something; from *that* point of view, we have failed to express a proposition at all (in this technical sense)—and so (from *that* point of view), we can be taken to have uttered something no better than a meaningless noise. I should stress this: Given the semantic purposes that we've determined certain idioms in the vernacular (and therefore the sentences in the vernacular containing those idioms) to have, we've determined that a certain subset of those sentences can be left out of account; proxies for them needn't appear in our regimentation. This analysis and the corresponding notion of "proposition" that I've offered here will be put to specific use with respect to truth in 4.8.[16]

4.4 Regimenting Truth: Self-Reference in the Vernacular

The treatment of AU-quantifiers in chapter 3 was predicative: The sentences of L_1 that aren't in L_0 aren't themselves *explicitly* in the domain of the AU-quantifiers of L_1. This makes natural a predicative AU-*hierarchy* to handle the regimentation of iterations of "true" in the vernacular, e.g., "Every sentence in Box A is true," where Box A contains the sentence, " 'John is running' is true." The AU-hierarchy is similar to the Tarskian one—at least insofar as both handle this class of truth-endorsements pretty much the same way.[17] Do we

---

[16] One other point should be made: Since the notion of "failing to express a proposition"—as I construe it—operates at the interface of the vernacular and its regimentation, adapting such a regimentation (and its proprietary notion of proposition) needn't require the adoption of truth-value gaps, three-valued logic, etc., in the regimentation; nor does it require attributing truth-value gaps, three-valued logic, etc., to the sentences of the vernacular—not, at least, on the grounds that certain sentences of the vernacular fail to express propositions in this proprietary sense. Instead the class of what's (grammatically speaking) expressible in the vernacular is *deliberately mismatched* with the class of what's (grammatically speaking) expressible in the regimentation of that vernacular. If the claim were instead that the regimentation was an (admittedly idealized) *empirical model* of the vernacular (see, e.g., Gupta 1984, 177–79)—and that the sentences in it were supposed to replicate the semantic properties of sentences from the vernacular, things would be different: We would *have to* introduce truth-value gaps, a three-valued logic, or something to that effect.

[17] However: The AU-hierarchy is much simpler, mathematically speaking, than the Tarskian hierarchy. Kripke (1984, 60–61) raises the technical issue of the transfinite iteration of the Tarskian hierarchy; such iteration is sensitive to exactly how previous truth predicates are coded into truth

need anything *more* to capture the blind-endorsement practices of ordinary language (that we *want* to capture)?[18]

My answer is a (tentative) "no," and I'll spend the remainder of this chapter motivating it. Kripke (1984, esp. 57–63) raises several powerful objections against predicative approaches to truth—his objections are specifically directed to what he describes as an "orthodox approach."[19] I'll look at these in 4.7 and 4.8. Apart from them, there are other motivations in the literature prior to Kripke's watershed article, and after, for analyzing truth-endorsement practices in the vernacular via a nonpredicative approach to truth. I'm not impugning the substantial mathematical interest such approaches have—at issue is their *philosophical value*.

Before undertaking all this, however, I must first waylay a possible misunderstanding: The predicative approach to AU-quantification *doesn't* rule out self-reference. As A. R. Anderson (1970, esp. 8–11) points out,[20] one can't, in any case, stipulate the elimination of self-reference from languages, formal or otherwise. And the predicative AU-hierarchy *doesn't*. Apart from the fact that self-reference can be present in the given language that AU-quantifiers are defined *on*—in the sense that the language of AU-quantification itself may be codable in such a language, nothing prevents the presence of *other* quantifiers defined on the language along with AU-quantifiers (or already existing within it) that directly allow self-reference.[21] So the issue isn't whether an approach to AU-quantifiers that rules out self-reference suffices for truth-endorsement practices in the vernacular; the issue is whether the predicative approach to AU-quantifiers suffices for the truth-endorsement practices in the vernacular that are *worth retaining*.

4.5 Semantic Completeness and Semantics as Motives for Impredicative Approaches to Truth

One motive for impredicative approaches to truth arises from the appearance that natural languages have of being "semantically complete." The particular

---

predicates at limit ordinals. I don't know the current status of the problem, but in any case, it *doesn't* arise for the AU-hierarchy because, unlike in the Tarski hierarchy, truth-endorsement is separate from a characterization of the syntax of the sentences so endorsed.

[18] This is *not* the technical question of whether impredicative approaches to AU-quantifiers are available; they are. Kripke and Gupta/Herzberger approaches to the truth predicate can be easily adapted to quantifiers, and in particular, to AU-quantifiers. The question is a different one: Is there evidence that such impredicative approaches are needed *at all*, given the needs of truth-endorsement practices in ordinary language?

[19] Kripke (1984, 58 n. 9) writes: "By an 'orthodox approach', I mean any approach that works within classical quantification theory and requires all predicates to be totally defined on the range of the variables"—but the objections raised easily apply to predicative versions of the AU-approach.

[20] Also see Kripke 1984, 58 n. 9.

[21] As it's been repeatedly noted, many examples of self-reference are benign in that the logical resources they're based on needn't give rise to either paradox or, more broadly, truth-value ungroundedness, e.g.: "This sentence has five words." These sorts of examples, of course, don't involve truth-endorsement, and don't need regimentation by AU-quantifiers.

form the concern takes in the context of our truth-endorsement practices is our ability to attribute truth or falsity to certain (large) classes of sentences in our language—to, for example, *every* sentence including, of course, the very sentence so attributing truth or falsity. One issue is whether, in fact, the vernacular *really does* allow such a thing. This question can be left aside. If the language gives the appearance that it allows such an ability, it may do so for reasons that force it to be inconsistent;[22] but in any case that's, of course, an empirical question about the vernacular. From the regimentalist's point of view, the important question is: Do we need a capacity to attribute truth and falsity to, say, *any* class of sentences *whatsoever*?

One may read such a need off of the practice of blind endorsement because one could think: *Any* description (of a set of sentences) can be used to characterize a set of sentences we want to endorse or deny. But this is too general a characterization of the function of blind endorsement. What's at issue—especially since we already know that impredicative devices that enable blind endorsement lead to paradox—is whether predicative devices, e.g., a predicative AU-hierarchy, leads to an inability to express proxies for sentences in the vernacular that speakers *really need*. As I said, I'm going to claim that the answer is "no."[23] My answer is tentative because a case-by-case study of specific sorts of blind endorsements that can be made in the vernacular, but are absent (if one adopts a predicative approach to truth-endorsements), is needed. (I do this in 4.8.) I want to stress again that the mere fact that a sort of (impredicative) blind endorsement exists in the vernacular is no argument—by itself—that regimentation needs to take account of it.[24]

It may be thought that semantic characterizations, complete in this sense, of the "truth conditions" of *all* the sentences in a language (including the very sentences *in* the semantic characterization) are especially valuable. But why? If empirical characterization of the semantic properties of natural languages is at issue, it's hard to see why it's required that such semantic properties be characterized within the language itself.[25] And if the issue is the semantic characterizations of the regimentations constructed for the normative purposes I've stressed in 4.1, again, it's hard to see why global (self-referential) semantic characterizations are required.[26]

---

[22] There are *very* good reasons to think natural languages *are* inconsistent, as I argue in chapter 5.

[23] Gupta (1984, 229) writes of such sentences—not just paradoxical ones, but other sorts of "ungrounded sentences" like truthtellers—that "from the ordinary viewpoint these are the 'don't care' cases."

[24] What motivates impredicative set theory, for example, is its value for constructions of set-theoretical ersatz for classical mathematical objects (see Kleene 1971, e.g., 42–43, for an accessible discussion of this). So, similarly, one wants to examine the specifics of truth-endorsement practices in order to see if there are motivations *in that practice* for impredicative truth-endorsements.

[25] After all, every other science allows itself to extend natural language in whatever way it needs to ply its trade. Why should semantics be different?

[26] The worry may be: *We'll never finish*. After all, we'd also like a semantic characterization of the (meta)language in which the semantic characterization is carried out. Okay, that can be supplied too; but why must everything be done *at once*? The thought of hierarchies of semantic principles seems to panic some, but that's hardly an *argument*.

4.6 *Diagnosing Ills in the Vernacular by Means of the Liar*

Another motivation for impredicative approaches to truth-endorsement also arises from a direct concern with the vernacular. This is the idea that the presence of paradox calls for an *explanation*: There is something to diagnose, something to discover about natural language that explains why paradoxes can arise or what they (really) are attributable to.[27]

It's unclear what a successful diagnosis is supposed to look like in this context. The resolution of paradoxes (when they are to be had) takes the following form: (i) paradoxes are parsed as contradictory sets of assumptions, and (ii) one of those assumptions is a hitherto unrecognized false principle that must now be rejected. Unfortunately, something like this—which is reasonable *given an antecedent subject matter that the assumptions are about*—can impel the thought (in the context of the Liar) that diagnosing paradoxes is teasing out an assumption to be exhibited as rejected by a reductio. And this, in turn, leads to something weird. Consider the sentence: (1) (1) is false. Imagine the following (purported) presupposition: (1) *refers to* "(1) is false," and consider this argument:

(a) (1) refers to "(1) is false."   Assumption.
(b) (1) is true if and only if (1) is   {use of standard Liar reasoning}
false.
(c) (1) doesn't refer to "(1) is false."   Reductio on (a) and (b).[28]

Or, consider *this* presupposition: (1) is either true or false. Again we find:

(a) (1) is either true or false.   Assumption.
(b) (1) is true if and only if (1) is   {use of standard Liar reasoning}
false.
(c) (1) is neither true nor false.   Reductio on (a) and (b).

I'm *not* claiming that one or another solution to the paradox that might be offered this way won't be interesting or won't be something we might eventually adopt; rather, the claim is that the intuitions generated by ordinary uses of reductio ad absurdum give the impression that there is something to explain about paradoxes, and that when an assumption that leads to the paradox is rejected, we've indeed explained how the paradox arises (as opposed to the very different methodological suggestion that nothing has been *explained*—that's the wrong model—but rather, we're exploring whether the "rejected" assumption is one we can or should live without).[29]

---

[27] The suggestion that semantic paradoxes are open to "diagnosis" seems to have originated with Chihara 1979. Of course, he was introducing a bit of jargon (which subsequently caught on) for something philosophers were previously clearly engaged in—as he admits.

[28] I explored this "solution" to paradoxes in Azzouni 1991. This isn't to suggest I motivated it by means of a reductio.

[29] In pointing out that such reductios are uninformative—beyond, of course, their exhibiting the *inconsistency* of a set of sentences—I'm *not* suggesting that philosophers generally don't realize this (for Quine and Popper, for example, this very insight is a joint substantial plank of their respective approaches to philosophy of science). More specifically, the most common official attitude

Again, it's important to separate the project of regimentation from the empirical study of language, and both of these from the project of determining the principles governing a subject matter. In the third case, a "paradox," an inconsistency in the set of principles governing a subject matter involves (we presume) a mistake *on our part*. We've wrongly formulated the principles governing that subject matter, and we must find out which one(s) should be rejected. Reductio ad absurdum only indicates that the principles, *as a group*, are problematical, and not which ones should be rejected. The (and this is where diagnosis, properly, arises) latter is determined by our understanding and study of the subject matter in question. Regimentation is a different matter entirely. Here, as I've said before, we're out to construct idioms that replicate functions of ordinary language items (that we want to retain), and it's a matter of sheer mathematics which ones have which properties. Again, there is nothing to diagnose; inconsistency is sheerly a (mathematical) fact about certain formal systems. Last, there is the empirical question of which syntactic (and semantic) principles are actually in use among speakers of a natural language. Here too there is nothing to diagnose because if the principles speakers use are shown (empirically) to be inconsistent, *then it just follows that the principles speakers use are inconsistent*. We can explain *why* an inconsistent set of principles are so used, e.g., what it is about the evolution of language that caused speakers to converge upon an inconsistent set of principles, and how they could retain such a set of principles *despite* an official norm *against* inconsistency, but *that* doesn't involve diagnosis as Martin or Chihara mean it. Rather, it seems from the foregoing that nothing involving diagnosis (in their sense) exists.

4.7 Kripke's First Objection to "Orthodox Approaches"

Let's turn to the objections Kripke offers against "orthodox approaches," objections, if sustained, that apply to the predicative AU approach as well.

---

toward semantic paradoxes is that there are many approaches and that one must delicately evaluate the vices of competing systems; see, e.g., Feferman 1982. Indeed, much earlier, Martin (1970) writes: "A solution [to the Liar] consists in convincing ourselves that at least one of the assumptions that led to the contradiction is after all not so plausible. Obviously if our own move in trying to remove the plausibility of a particular assumption is to treat the argument to contradiction as a *reductio*, as though it proves that the assumption in question is false, we have failed entirely." He adds: "What is wanted, ideally, is the uncovering, the making explicit, of some rulelike features of our language which when considered carefully have the effect of blocking at least one of the assumptions of the argument; if not actually showing an assumption to be false, at least casting doubt upon it." Martin, near as I can tell, doesn't reveal what tools we're to use to *empirically determine* that the targeted assumption is *not* operative among speakers.

The reader may be wondering why I bother to make this point if philosophers generally *are* aware that such reductios are uninformative. Well, they're aware in *some* contexts but not in others. I'll allude to two examples: One motivation Herzberger (1982) offers for the (otherwise purely mathematical) study of how the truth-value revision process he uses gives rise to cycles among pathological sentences is the diagnostic one (see p. 135, in particular). For reasons I give momentarily, it's hard to see how any diagnostic aim can be satisfied in this way. For a criticism of a second example from the revisionary literature, see Azzouni 1995.

Here's the first: Regimenting the vernacular via predicative AU-quantifiers, like regimentation via the standard Tarski hierarchy, seems to impose subscripted idioms (truth predicates or AU-quantifiers), each with its own level.[30] Kripke (1984, 58–60) writes:

> [This approach takes] the ordinary notion of truth [to be] systematically ambiguous: "level" in a particular occurrence is determined by the context of the utterance and the intentions of the speaker. The notion of differing truth predicates, each with its own level, seems to correspond to the following intuitive idea. . . . First, we make various utterances, such as "snow is white", which do not involve the notion of truth. We then attribute truth values to these, using a predicate "$true_1$". . . . We can then form a predicate "$true_2$", . . . and so on. We may assume that, on each occasion of utterance, when a given speaker uses the word "true", he attaches an implicit subscript to it, which increases as . . . he goes higher and higher in his own Tarski hierarchy.

Kripke goes on to say:

> This picture seems unfaithful to the facts. If someone makes an utterance [such as "Most (i.e., a majority) of Nixon's assertions about Watergate are false"] . . . he does *not* attach a subscript, explicit or implicit, to his utterance of "false" which determines "the level of language" on which he speaks. . . . [The problem is that o]rdinarily . . . a speaker *has no way of knowing the "levels" of Nixon's relevant utterances*. . . . The idea that a statement such as (4) ["All of Nixon's utterances about Watergate are false"] should, in its normal uses, have a "level" is intuitively convincing. It is . . . equally intuitively obvious that the "level" of (4) should not depend on the form of (4) alone (as would be the case if "false"—or, perhaps, "utterance"—were assigned explicit subscripts), nor should it be assigned in advance by the speaker, but rather its level should depend on the empirical facts about what Nixon has uttered.

This objection is eluded if the hierarchy is supposed to be a regimentation. For then it's not required that the ordinary truth predicate, say, *have* subscripts. All that's required is that a regimentation that subscripts truth predicates, and locates proxies of ordinary-language sentences (on the basis of those subscripts) in the hierarchy, be the best artificial approach (known to date) for facilitating truth-endorsements. In particular, speakers needn't be taken to have any intentions whatsoever about the location of their truth-endorsements in a hierarchy; and it can, of course, be an empirical matter what level the proxy of an ordinary sentence is assigned in the Tarski hierarchy, and one based (of course) on what sentences that sentence is about.[31]

---

[30] The "ground floor" AU-quantifiers regiment truth-endorsements as sets of sentences which themselves don't involve truth-endorsements. The next level includes the first collection of sentences plus truth (and falsity) endorsements to sentences (and groups of such) in the first collection. And so on.

[31] It's clear that Kripke (1984) was not directing his objections toward a regimentation view. I have to say again, though, that the formal tools used in the Tarskian, Kripkean, and for that matter, in *all* the approaches taken in this literature, aren't ones that have been (or, I think, can be) justified *if one is actually engaged in an empirical study of natural languages*.

4.8 Kripke's Second Objection to "Orthodox Approaches"

Kripke's second objection is far more important. As he (1984, 60) points out (and he is, I think, the first to point this out), there are examples of truth-endorsements *in the vernacular* that seem to resist a predicative approach and nevertheless are (intuitively) assigned truth values.[32]

Kripke's example is the pair of statements uttered by Dean and Nixon respectively:

(1) All of Nixon's utterances about Watergate are false.

(2) Everything Dean says about Watergate is false.

Both speakers—somewhat megalomaniacally (but quite within charac-ter)—want to include all of each other's statements within the scope of their quantifiers—and this makes location of either statement in a predica-tive hierarchy perilous. Unfortunately, the hope of simply ruling out such claims as unsuccessful (as not expressing propositions, in the sense of 4.3) seems to founder on the naturalness of assigning both (1) and (2) truth values under certain empirical circumstances. Imagine (as Kripke suggests) Dean has uttered at least one true statement about Watergate (other than (1)); then on these grounds alone we take (2) to be false. *And* if everything else Nixon ever said about Watergate is false then (1) (we naturally think) is *true*.[33]

Another neat example comes from Gupta (1984, 210). He introduces his example with a significant remark: "There are types of reasoning that we allow in everyday discourse . . ."

A says:

      (a1) Two plus two is three. (false)
      (a2) Snow is always black. (false)
      (a3) Everything B says is true. (----)
      (a4) Ten is a prime number. (false)
      (a5) Something B says is not true. (----)

B says:

      (b1) One plus one is two. (true)
      (b2) My name is B. (true)
      (b3) Snow is sometimes white. (true)
      (b4) At most one thing A says is true. (----)

---

[32] In a sense, all the forthcoming examples are generalizations of "All Cretans are liars"; this statement, recall, when uttered by a Cretan, is paradoxical if and only if every *other* statement uttered by a Cretan is false. Otherwise we're quite willing to treat this statement as *unremarkably false*.

[33] As Kripke also points out, it's easy to change the empirical circumstances to reverse the truth values for (1) and (2); it's also easy to construct other sorts of examples.

Gupta points out that it's very natural to reason as follows: Since (a3) and (a5) contradict each other, (b4) is true. Therefore (a3) is true and (a5) is false.

The importance of these examples for *our* purposes is this.[34] Even if one finds the project of studying *paradoxical* sentences of no interest—at least when it comes to evaluating our blind-endorsement needs in the vernacular—and even if one thinks our ability (in the vernacular) to construct paradoxical sentences *doesn't* show anything significant about the adequacy of predicative approaches to truth-endorsements (even if, that is, one is comfortable leaving paradoxical sentences out of one's regimentation of our truth-endorsement practices, and even if, in doing so, one concedes that it's an empirical matter out of the speaker's hands whether the truth-endorsement a speaker makes *does* express a paradox or not), these examples pose a problem precisely because they are *not* examples of paradoxes: They are cases of blind endorsement that we're intuitively inclined to treat as successful (in the sense that truth values—truth *or* falsity) are assigned to all the sentences in question. This seems to show not only that it's an empirical matter, out of the hands of speakers, whether their truth-endorsements *are* successful or not (this is the motivation for Kripke's oft-repeated assertion that such truth-endorsements should be "allowed to find their own level") but, more important, that any regimentation of truth-endorsement which leaves us without resources for directly replicating nonpredicative kinds of blind-endorsement reasoning fails to capture a large class of blind endorsements which speakers routinely make (since, after all, one can assume that speakers routinely utter successful truth-endorsements that under other empirical circumstances would have proven paradoxical, and thus which resist predicative treatment).[35]

Just because truth-endorsements of speakers go sour in ways speakers can't anticipate, *doesn't mean* that the ways such sentences do go sour need be captured by regimentation. After all, the common response to paradoxes (when presented at parties during, say, the course of ordinary chatter) is to move on (after, perhaps, a barely polite laugh or two). There's *no* evidence that ordinary speakers *do* anything when faced with sentences that *are* paradoxical or otherwise truth-value defective (e.g., truthtellers) that shows that there is something special about their reasoning with such items that needs capture by regimentation. The ordinary person's practice of ignoring such sentences is

[34] Kripke and Gupta have *different* purposes, of course. Kripke (1984) takes his examples to show that a predicative approach to truth won't work. Gupta (1984) takes his example(s) to illustrate sorts of reasoning in the vernacular that Kripke's approach has trouble accommodating.

[35] A technical point. Predicative (or hierarchical) approaches, as I understand them, assign levels to sentences based only on which sentences they refer to and quantify over. (4) and (5) are immune to a level assignment on these grounds alone. Kripke's approach is to assign such sentences "levels" (stages at which they're assigned truth values) based, in addition, on the *truth values* assigned to the sentences they refer to and quantify over.

reasonably imitated by the regimentalist (that is, by treating such sentences as failing to express propositions, in the proprietary sense of 4.3).[36]

It's the fact that under certain empirical circumstances a sentence that resists predicative treatment *is nevertheless assigned specific truth values* that causes a problem. But what's important isn't merely that speakers naturally assign truth values to such sentences; the critical issue is whether the fact that they do so is significant enough to our blind-endorsement practices to require us to replicate such sentences—and the truth values assigned—in our regimentation.

One important consideration suggests that it *isn't* significant. Recall the point made earlier about attempts to "diagnose" paradoxes. The logical principles we've imbibed along with our language are ones it's natural for us to employ *anywhere we can*.[37] And what this means is that even with sentences that, strictly speaking, we have no genuine need for (given the purposes our truth-endorsing practices satisfy), we may find ourselves with logical intuitions resulting from the (rote) application of those principles.

Consider the lying Cretan again (recall note 32). We *can* reason our way a priori to the empirical claim that some Cretan or other must not have lied; for in this way we avoid paradox. Ordinary speakers carefully restrain their otherwise exuberant application of logical principles, such as reductio ad absurdum, precisely because such principles otherwise lead to the assignment of truth values to empirical claims in which no such inferences are (empirically) justified. But the only difference between this example and the ones Kripke and Gupta offer is that since none of the claims made in the latter examples are in danger of leading to empirical refutation (because they're ungrounded!), we allow ourselves to rest with the intuitive assignment of truth values dictated by those principles, even though the kind of reasoning involved *differs in no significant way, logically speaking, from the (unjustifiable) inference to the existence of a lying Cretan.*

And *this* shows that the (rote) reasoning exhibited in examples such as Kripke's and Gupta's *isn't* enough—all on its own—to fault predicative approaches that simply bar such examples out of court. Since these are the only sorts of examples that anyone has ever offered to show that reasoning in the vernacular about "ungrounded" sentences should be taken seriously, I draw the conclusion that the regimenter can ignore impredicative truth-endorsements altogether without, as a result, failing to include instances of (otherwise already present) significant and valuable inference patterns. Although it's an empirical matter—out of the speaker's hands—whether one of these has been uttered inadvertently, characterizing such sentences (distinguishing impredicative from predicative uses of the truth predicate in the vernacular) can be executed

---

[36] Again, this is *not* to exclude the purely empirical study of establishing what it is about the semantic rules that *actually govern* natural languages that allows such sentences into the vernacular in the first place.

[37] That is, our *logical principles* are topic neutral. One reason for this is confirmation holism. I intend to take up the topic neutrality of logic in some detail at another time.

without engaging in Kripke or Gupta/Herzberger style constructions, because the distinction, although semantic, doesn't turn on the assignment of truth values.[38]

4.9 Other Misgivings about Hierarchal Approaches to Truth

More vaguely put (but still rhetorically compelling) objections to hierarchy approaches occur in the literature regularly. Echoing this traditional source of discomfort, Simmons (1999, 475) writes: "The English predicate 'true' is divided into infinitely many distinct predicates, and English itself is stratified into a hierarchy of distinct languages. This seems, as Russell once put it, 'harsh and highly artificial.'"

This concern may be ascribable, to some extent, to the preconception that hierarchy views require attributing to competent speakers an appreciation of the location of sentences in the hierarchy. If, as I've argued in 4.7, such views

---

[38] There is another objection to these approaches, but since it's not as central to the topic of this chapter, I'll confine it to a note. Crucial to the intuitions elicited by the Kripke/Gupta examples isn't just what truth values *are* assigned to such sentences but *how* we reason ourselves *to* these truth values. Although Kripke's formal construction supplies the right truth values for the examples he discusses, and although Gupta's formal construction supplies the right truth values for the examples *he* discusses, neither approach assigns such truth values in anything like the way ordinary reasoning does it: Neither step-by-step reasoning from sentences at lower levels to higher levels (as in Kripke's approach) is at work—as Gupta's examples show—nor is revisionary reasoning present, as Gupta's approach requires. Despite both philosophers' (with hedging on Kripke's part) treating their formal techniques as capturing, to some extent, the informal notion of truth itself (as opposed to, at best, an extensionally equivalent formal analogue), this wouldn't be an insurmountable problem if such approaches assigned the *same* truth values to pathological sentences as ordinary intuition does. Unfortunately, all such approaches breed *artifacts* because of how their impredicative models are constructed: Sentences, that is, are given truth values at variance to ordinary intuition—for example, sentences ordinary intuition regards as pathological are stably true or false (in the Gupta-Herzberger approach). (See Azzouni 1995 for some details on this.) This is good evidence that what's involved in the intuitive assignment of truth values in the kinds of "circular reasoning" examples given by Kripke and Gupta is the extension of already-in-place ordinary practices of reasoning to "ungrounded sentences," and not something more drastic—such as a "revision concept of truth."

Another indication of this very same point is that both Kripke's and Gupta/Herzberger's approaches distinguish sentences according to their sort of pathology ("paradoxical," "intrinsically paradoxical"...). They *do not* illuminate or even indicate what sort of reasoning may be appropriately applied to such sentences—thus they ratify the "don't care" attitude taken toward them by practitioners in the vernacular. (See note 23.)

I hope no reader mistakenly thinks I'm suggesting that this faults *in any way* the examples of self-referential reasoning that arise in technical areas; these aren't examples of blind truth-endorsement that require nonpredicative approaches: These are syntactic results based on the existence (or not) of certain sorts of predicates in certain systems. Even in the case of Gödel's theorem, which is often popularly described as a proof that there are true sentences which aren't provable, what's going on is a proof of the syntactic incompleteness of a certain class of formal systems; and, subject to the co-referentiality of (some of) the terms in these systems, the attribution of truth to sentences derivable in some of these formal systems but not in others. (See Azzouni 1994, part II, § 7, for details on this. Also see 9.9.)

don't require any such thing, then the mere fact that the truth predicate in English doesn't intuitively seem scattered among a hierarchy of languages is no objection to these approaches—nor, of course, is their apparently harsh and artificial quality. Another source of the ill ease quoted above may be, as we've seen before, the clash of seriously different *projects*: a genuine empirical study of ordinary language on the one hand versus regimentation on the other.

4.10 Concluding Remarks

There's nothing simpler than the (simple) liar paradox: It can be understood by a child. And, in a sense, there's nothing simpler (as well) about how reductio ad absurdum can snooker us into "solutions" of such a thing. (The intense but naïve delight one feels when first thinking, for example, "that's it!—not *every* sentence is true or false.") But really, the "diagnosis" is this: *There's nothing to diagnose.* Certain languages—most likely our natural ones, as I argue in chapter 5—allow *ineliminable* inconsistent sentences via syntactic and referential principles that aren't hard to understand. If there were a simple way to replace such languages with ones that *don't* imply such contradictions—to change their syntactic or referential principles in some way—and yet have a resulting language as expressive as natural languages (at least with respect to those parts of natural language being modeled), we'd do so immediately. This wouldn't be a diagnosis of *anything*. The intuitive gloss on reductio ad absurdum that, using it, one unearths a false assumption about a subject matter is just *wrong*. Rather, we would have discovered the (mathematical) existence of such a language. We could then regiment our own language in it: proxy (some of) the sentences of the vernacular in that artificial discovery, and leave for another time the difficult and subtle empirical question of whether our natural languages are, in some way, like this artificial discovery, or whether, as they appear, natural languages are simply inconsistent. Unfortunately, no such desirable artificial language exists.[39] Luckily, we can circumscribe those cases of truth-endorsement that we need—regiment them in a predicative hierarchy, and discard the rest. That it's an empirical question whether certain (chancy) sentences we've uttered *are* ones that can be so regimented, and thus treated as successfully expressing a proposition, or not, is yet one more fact of life that we'd better accommodate ourselves to.[40]

The state of play is this: To some extent I've justified the use of AU-quantifiers as a regimentation of our truth-endorsement practices in the

---

[39] Apart from all the work exploring what sorts of languages—that allow various sorts of self-reference—are possible, there is almost a proof of this fact, the existence of which can be recognized via the very common criticism offered to solutions of paradoxes that impose contextual levels on paradoxical sentences, e.g., C. Parsons 1984 or Burge 1984, so that such a sentence (on a use) fails to say of itself (on that use) that it is false but succeeds only in saying of some other use (of that sentence) that *it* is false: This is that such an approach fails to allow the paradoxical sentence to say what's intended. But, of course, if it did *that*, it would still be paradoxical.

[40] We'd have to live with it regardless of our approach to paradoxes—this is one of the most important lessons of Kripke 1984.

vernacular, and I've tried to justify, more specifically, a predicative construction of AU-quantification to those practices. This still leaves the question, raised in 3.5, of exactly how truth-endorsement practices in ordinary languages work, specifically, how in the absence of AU-quantification in the vernacular the truth predicate still manages to be a transcendent one. I'll eventually sketch a story of how this goes (see the conclusion to part I), but first I want to establish a much more dramatic fact about natural languages: that they are, in a perfectly prosaic sense, inconsistent. Establishing *this* is the task of the next chapter.

5

The Inconsistency of
Natural Languages

5.1 The Revenge Problem

Consider the sentence:

(1) (1) is false.

Because of simple naming conventions that everyone knows, (1) implies (by ordinary reasoning that everyone can easily execute):

(2) (1) is false if and only if (1) is true.

A quick response to (2) (and one, by the way, that non-philosophers often think of) is to claim that (1) (contrary to first impressions) isn't "meaningful," or "doesn't express a statement"—or a "proposition"; more sophisticated views that come to the same thing (at least insofar as they prima facie disable the repulsive (2)) are that (1) is "ungrounded," or is "truth valueless," and so on.

But this precipitates a second paradox: Let "**S**" stand for your favorite characterization of (1) that allows (2)'s avoidance. Then by reasoning almost as simple as the reasoning that leads to (2) (and reasoning, by the way, equally accessible to the ordinary person), we find that:

(3) (3) is false or (3) is **S**

implies

(4) (3) is true if and only if: Either (3) is false or (3) is **S**.

Priest (1987, 29) writes of strengthened liar paradoxes that they "are not really novel paradoxes, but merely manifestations of one and the same problem, suitable to different contexts." For the derivation of (2) from (1) presupposes a division of sentences into the true and the false; and the response to (2) is to subdivide the territory differently: into the true, the false, and the **S**; but the false and the **S** can be bundled into, as Priest puts it, "the Rest," where the Rest strictly *includes* the false and strictly *excludes* the true. And then, as we've seen, it's easy to formulate a new paradox in terms of the true and the Rest. Thus, a natural solution—one anyway, that has occurred to many philosophers—to strengthened liars is to ban (Priest 1987, 29) "the expressibility of certain key concepts (truth, Value gaps, stable truth etc.) from the language." And this "requires the paradox solver to insist that she herself is talking in a language different from the one for which the semantics are being offered (the 'metalanguage')."

Although Priest (as quoted above) considers strengthened liars to be twisted versions of the ordinary liar, it's clear that the introduction into the formulation of the strengthened liar of new (and usually fairly technical) vocabulary makes the not-expressible-in-the-language-of-the-original-paradox response to strengthened liars an option. This has engendered disagreement about the expressive resources of natural languages. Kripke (1984, 80 n. 34) writes:

> Such semantical notions as "grounded", "paradoxical", etc. belong to the metalanguage. This situation seems to me to be intuitively acceptable; in contrast to the notion of truth, none of these notions is to be found in natural language in its pristine purity, before philosophers reflect on its semantics (in particular, the semantic paradoxes).

Burge (1984, 88 n. 8), however, responds: "I see no interesting or clear distinction between terms reflectively introduced into natural language for unchallenged explanatory purposes and terms that slip in by other means." Priest, expressing similar sentiments, writes that the notions necessary for the formulation of strengthened liar paradoxes (1987, 20) "obviously are expressible in English."[1]

This is just the sort of debate that looks philosophically irresolvable.[2] Perhaps, in this case, this is because English, like every natural language (especially these days) swells daily with new words; despite this, native speakers of English don't feel required either to change the name of the language or, perhaps more

---

[1] Priest *means* it: While discussing the hierarchical view he claims that the notion of a sentence's rank is so expressible.

[2] Consider, for example, my copy of Felsager 1998. Is it written in *English*? Well, in some sense it surely must be. Opening it at random, I found myself at page 234, where I read: "Since we are not interested in the elevated energy states, we need not evaluate the total path-cum-trace integral." I *guess* that's English (although I'm sure lots of native speakers might disagree); after all, the author, in his acknowledgments, writes (vi): "Helge Kastrup Olesen looked over a preliminary version of the manuscript and taught me a lot about English grammar." On the other hand, this is precisely the sort of book that a non-English-speaking student (an appropriately trained one, I mean!) would manage to get rather a lot out of—and despite, say, being (in the traditional sense) utterly ignorant of English.

important, even to worry about the identity conditions on languages, and whether they've been violated by particularly numerous or far-reaching changes.[3]

The real point of contention between Kripke and his opponents, however, is about the expressive strength, whatever it is, of natural languages, and not about whether, as the particulars of the exchange between Kripke and Burge suggest, certain terminology belongs (or not) to natural languages (in their "pristine purity"). And this is because one can allow that the terminology necessary to the expression of a strengthened liar is present in natural languages, and yet claim that nevertheless strengthened liars don't arise because such terminology is present only in a stratified way; "expresses a proposition" and so on apply not to sentences containing themselves but only to designated portions of ordinary language free of the terms in question.

What can help motivate the view that, at least in principle, it's possible for the idioms used to express a strengthened liar (such as "doesn't express a proposition," "is gappy," etc.) to be treated differently—in natural languages—from the way "true" is treated, is the recognition that the role of the truth predicate, in ordinary language, is *not* a genuinely semantic one but serves only to facilitate semantic ascent and descent in blind endorsements. As I argued in 2.1, this makes Tarski's conjoining of a theory of the semantics of a class of formal languages with a theory of the truth idiom something of an accident of the particulars of the history of the development of formalized languages. In this sense, therefore, we can honor Kripke's point (quoted above): We can argue that, in any case, the truth idiom isn't wedded to other syntactic and semantic notions— even if there is no convincing argument that such notions, being technical and sophisticated, aren't part of natural language in its "pristine purity."

As appealing as I obviously find the line of thought I've just explored to be, I suspect it won't work. The problem is that, as I stressed at the very beginning of this section, access to strengthened liars is available to ordinary speakers, and so is access to the reasoning that establishes their paradoxicality. Although intuitions of this sort are never the final word, evidentially speaking, still it seems clear that if we press the hierarchical view—*in natural languages*—with respect to the more rarified semantic idioms used to construct strengthened liars, we're doing so only because of our desperation to escape the paradoxical conclusion (4), and not because of empirical evidence that such terms, as they are employed in natural languages, *really do* apply hierarchically.

5.2 Katz's Solution to the Liar Paradox

As I understand Katz's solution to the liar's paradox, it goes like this.[4] There are two kinds of propositions, intensional ones and extensional ones. (1)

---

[3] I'm not, of course, speaking of *French*, which in this respect is hardly a *natural* language.

[4] I draw my understanding of his approach from 3.6 and 3.7 of his 2004, and to a lesser extent, from Katz 1972, 136–38.

expresses an intensional proposition, and thus (1) is meaningful in this technical sense; furthermore, that it's meaningful in this technical sense explains why it's meaningful to ordinary speakers. Intensional propositions are sense-bearing vehicles (more generally, bearers of "sense properties and relations"), but they are *not* truth-value bearing vehicles. Only extensional propositions bear truth values. Now an unproblematical sentence "John is running," presumably expresses both intensional and extensional propositions, that is, in addition to making sense, it also makes a statement that carries a (determinate) truth value, but (1) doesn't, and since it doesn't, it doesn't have a truth value (and so the reasoning that purports to establish (4) misguidedly presupposes (1) as open to—one or another—truth value).

Katz is clear that a sentence successfully expressing an intensional proposition turns on its having lexical items that are all meaningful terms of the language, *and* on there being no step in the compositional process yielding that sentence from these lexical items that violates a restriction on "the compositional combination of senses." What does a sentence making a statement, that is, expressing an extensional proposition, turn on? No doubt many factors are relevant.[5] But what's most pertinent for us is that the sentence be grounded,[6] which Katz (2004, 101–2) glosses as making "the statementhood of assertions about assertions depend on the reference of their terms not entering a loop within which reference cycles endlessly without terminating in an unproblematic truth or falsehood."

This formulation clearly treats groundedness as a necessary condition for statementhood, and consequently as a necessary condition for susceptibility to either truth or falsehood. But a finer characterization of what groundedness actually comes to (given the above description, anyway) is a little elusive. Kripkean groundedness is a function not just of which sentences refer to which, but also of what truth values they're susceptible to. This means that some sentences that can seem referentially ungrounded (if we just focus on referential loops, for example) aren't so on the Kripkean view.[7] On one interpretation, Katz's quoted position requires a sentence that expresses a statement to be amenable to location in a predicative hierarchy; that is, all referential chains among the sentences must be grounded; more generously (and taking seriously his citation of Kripke 1984), Katz's quote, especially the phrase "unproblematic truth or falsehood" might be interpreted as acknowledging a Kripke-style approach to groundedness. In either case, such a view categorizes strengthened liars, in particular those such as

---

[5] For example, that all its presuppositions be satisfied. Katz 1972 explicitly suggests that presuppositional phenomena are operating in liar paradoxes. But Katz 2004 instead describes his solution to the paradox in terms of an assumption mistakenly held by philosophers: that "Fregean intensionalism is the only form of intensionalism."

[6] Here Katz cites his own unpublished work with Herzberger (Herzberger and Katz 1967), as well as Herzberger 1970, and Kripke 1984.

[7] Recall Kripke's Nixon/Dean example from 4.8. Intuitively, these statements can be assigned truth values, despite the fact that, referentially speaking, they loop.

(5) (5) is false or (5) is ungrounded;

for (5) is, pretty clearly, ungrounded, and so, despite the fact that it makes sense, and despite the fact that its sense seems to indicate that it's *right* about its ungrounded status (and therefore paradoxical, if ungroundedness and truth exclude each other), the sentence nevertheless fails to make a statement. For that matter,

(6) (6) is ungrounded,

is ungrounded, and therefore fails to make a statement, even though it makes sense, and seems (intuitively) true.[8]

So the groundedness approach sits uncomfortably with ordinary intuition. And indeed, the objection to the groundedness approach, that we intuitively accept certain ungrounded sentences as true or false despite their being (in one or another technical sense) ungrounded, was raised long ago. Taking, for example, Kripke's minimal fixed points as good models for the grounded sentences, Gupta (1984, 210–11) gave examples, as we saw in 4.8, in which intuitively appealing truth-endorsements can be distributed among a group of sentences, although none of those sentences are given truth values in the minimal fixed points; he also noted that ungrounded sentences include more than paradoxical sentences, and even though such sentences, e.g.,

(7) (7) is true,

are without truth values, this shouldn't be the case with sentences such as

(8) "(7) is true" is true if and only if (7) is true.

Although (8) is intuitively ungrounded, that doesn't prevent its seeming clear that nevertheless (8) is true.

Katz (2004, 102) bites the bullet on this one; he writes:

> Of course, to endorse the groundedness approach is not to say that any present theory of groundedness is everything we want in a satisfactory solution to the Epimenidean paradoxes. There is, as far as I can see, no reason to think that we will not at some point achieve a fully satisfactory theory, but, of course, there is no guarantee. It may turn out that every groundedness condition we formulate has undesirable consequences to the effect that statements we think we can make cannot be consistently made.

---

[8] (6) methodologically challenges the proponent of groundedness. For that proponent (Katz in particular) takes ungroundedness to exclude truth or falsity. This is what drives (5) into having paradoxical status (rather than being simply true), and thus into gappy status, and (6) directly into gappy status (rather than being true). Ordinary intuition tempts one to take (6) (and (5), for that matter) to be true *and* ungrounded. But then ungroundedness fails to be coextensive with a failure to express a statement, and more generally, to be coextensive with gappiness. Usually—at this point—one mutters the usual bit about not being able to honor *all* the intuitions ordinary speakers have about self-referential phenomena, and offers the methodological maxim that one need honor only the most central such intuitions, or something like that. I say something about this suspicious maneuver in 5.4.

Unfortunately, as the foregoing indicates, there is precious little evidence, empirically speaking, to justify the claim that paradoxes, strengthened or otherwise, fail to make statements. On the other hand, there is good evidence that sentences genuinely missing presuppositions (e.g., the venerable "the present king of France is bald") are easily recognized by speakers as so missing them—in practice, for example, speakers *deflect* a request for a truth attribution to such sentences by pointing out the missing presupposition ("Er, maybe you haven't *heard*; there *is no* present king of France"). This is *not* something speakers do with liar paradoxes. Hence the genuine puzzlement that such paradoxes raise for *ordinary* speakers.

This is an important methodological point. A substantial theory of presupposition may pose puzzles for theorists of language—how exactly does presupposition work, and how does it function (exactly) in the compositional attribution of meaning to sentences?—but it hardly poses puzzles for ordinary speakers (they simply operate with presuppositions in a way that theorists are supposed to *discover*). That liar paradoxes create puzzles for ordinary speakers warns us that there is something fishy, not in the subtle sense that there is a presupposition or false assumption that speakers have mistakenly adopted, and that the clever philosopher can ferret out for them, but in the sense that there is something fishy *in the practice itself.*

5.3 Opposition to the Suggestion That Natural Languages Are Inconsistent

Katz, as he makes clear, is motivated in part to pursue his particular route because he finds one of the alternatives intolerable. This is the suggestion, embraced by Priest (and, of course, by Tarski) that ordinary languages are inconsistent. Katz is hardly the first to be, well, *outraged* by this suggestion. Indeed, despite Tarski's stature, the *scorn* (over the years) heaped on this particular view of his is impressive.[9] Burge (1984, 83–84) writes:

> The best ground for dissatisfaction [with Tarski's view that natural language is hopelessly infected with contradiction] is that the notion of a natural language's harboring contradictions is based on an illegitimate assimilation of natural language to a semantic system. According to that assimilation, part of the nature of a "language" is a set of postulates that purport to be true by virtue of their meaning or are at least partially constitutive of that "language". Tarski thought that he had identified just such postulates in natural language as spawning inconsistency. But postulates are contained in theories that are promoted by people. Natural languages *per se* do not postulate or assert anything. What engenders paradox is a certain naïve

[9] Tarski (1944, 349) later reneges, but for a really interesting reason: Although everyday language prima facie looks inconsistent, "the case is not so simple. Our everyday language is certainly not one with an exactly specified structure.... Thus the problem of consistency has no exact meaning with respect to this language." As the reader will see, Tarski's dissatisfaction with his original formulation hasn't the same source as that of his detractors.

theory or conception of the natural concept of truth. It is the business of those interested in natural language to improve on it.

Katz (1972, 23 n. 6) expresses similar sentiments, and he opposes Herzberger (1970) when the latter suggests that language might fail to be "universal":

> We find that [Herzberger] has not shown any need for restricting the claim that full effability is the essential feature of natural languages. The appearance to the contrary depends on failing to make two distinctions, first, between a language and a theory, only the latter of which has postulates, statements put forth as the basic truths about some domain.[10]

Putnam (1975a, 73) makes what seems to be a similar objection:

> [The claim that all natural languages are inconsistent (because they are "semantically closed")] is false [in part] because only *theories* (systems of assertions) are inconsistent, and natural languages, e.g. English, are not theories. Someone speaking English may assert that there is a set of all those sets that do not contain themselves, or that all grammatical English declarative sentences are either true or false, but "English" does not assert these things![11]

It's hard to see why these considerations—despite the eminence of their sources—should have any force at all. Once we drop the view that ordinary-language practices—which include the logical principles that practitioners of a natural language have mastered as part of their acquiring their natural-language competence—are supposed to be explicitly postulated,[12] nothing stops the *possibility* that the practices in natural languages will imply (as, in fact, they clearly do) a contradictory grammatically well-formed sentence. After all, it's not that speakers, for example, need to be able to state a principle to the effect that any name can be used to refer to any sentence, regardless of what names appear in that sentence, or something like that. Nor is it required of speakers that they be able to formulate the appropriate generalizations that govern their truth-endorsement practices, or that they be able to formulate the logical principles needed to apply those generalizations to the specific sentences that speakers routinely apply them to. All that's needed is for speakers to naturally accept every step that leads to the contradiction, given a liar paradox. What happens *next* is also evidentially relevant: exactly how speakers, when faced with the resulting contradiction, *try to backtrack*, attempt to find a step or assumption that they can describe as a "mistake" (thus the "dialectic" that generates the strengthened liar).[13] It's up to theorists, of course, to formulate (correctly) the principles that speakers are (implicitly) relying on. But this is a

---

[10] I omit the second distinction, since it isn't germane.

[11] According to the footnote on page 70, this paper was read at Oxford in 1960, but it seems that it didn't appear in print until 1975. I don't know if Putnam expresses similar sentiments in earlier published work.

[12] A view that comes dangerously close to embarking on the infinite regress that Quine (1975c) warned us about.

[13] I'll have more to say about this important kind of evidence in 5.6.

different matter, and doesn't bear on whether the (implicit) principles speakers are helping themselves to are consistent.

Putnam (1975a) and Burge (1984) seem to presuppose—I'm speculating here—a notion of "language" in which grammaticality and "(semantic) postulation" can be cleanly separated, the former taken to be a property of languages proper, and the latter taken to be a property of something else (but what? Folk reasoning?) and where, furthermore, it's *obvious* (to the point of near embarrassment) that these two items are cleanly separable in the way that these philosophers want. But this *can't* be right, if only because how names may be assigned to sentences is only dubiously assigned to the "semantics" so understood. Rather, it's clear that how names can refer is part of what one *learns* when one learns a language, and so, the conclusion is that (at least implicitly) languages *do* make assumptions, and in just the same way that languages make demands about what is to be, and what isn't to be, grammatical.[14]

One can, of course, accept Burge's (1984) point that a *theory* of language might engender paradox for the same reason that one can accept the fact that *any* theory can engender "paradox"—because it has inconsistent assumptions that can appear disguised in one form or another.[15] But that's a different point, and one that doesn't affect the cogency of a theory of natural language—a consistent one, by the way!—correctly attributing inconsistency to those languages in the sense just specified: The implicit principles that speakers imbibe when they achieve competence in a language are inconsistent ones.

Katz, it might seem, has even less of a right to this line of thought about the impossibility of inconsistent languages since, although he too rails against an inappropriate assimilation of language to theory, he thinks nevertheless that languages have "analytic" entailments. Katz writes (2004, 95):

> Since the analytic sentences of a language owe their semantic status entirely to the structure of the language, the consistency of a language is a matter of the set of its analytic sentences not entailing a contradiction. Thus, for a natural language to be inconsistent, there would have to be a derivation exhibiting a contradiction among its analytic sentences. But, as Hans Herzberger points out (1965), since the conclusion of the derivation purports to be a contradiction logically derived from statements none of which, being analytic, can be false, the conclusion must itself be false. It is a *reductio* of the claim that there is any such derivation. Thus, natural languages must be consistent.[16]

Any such "reductio," of course, fails ("One person's modus ponens is another person's modus tollens."). Or, to put the point another way; one can't

---

[14] And these inconsistent assumptions can be made to look more like a "genuine" paradox than a mere contradiction if the inconsistent assumptions are tucked into inference rules, or what we might call, more generally, grammatical rules (e.g., that a name isn't restricted in what sentences it can be applied to), or into definitions. See, e.g., Chihara 1979, for illustrations of this.

[15] There is no mechanical test for inconsistency. This remark is so important that it's worth repeating.

[16] I draw the material Katz refers to here from Herzberger 1966, which appears in a revision and expansion of the 1964 issue, where the Herzberger article first occurred.

stipulate that the analytic sentences can't be false; not, anyway, if one simultaneously accords them an empirical role (e.g., "owing their semantic status
entirely to the structure of the language").[17]

5.4 The Inconsistency of Natural Languages

As I've already suggested, I think that it's not only possible for natural languages to be inconsistent, as Tarski (1983a) once suggested, but in fact the
case. This is, of course, an empirical claim, and so I hope the reader doesn't
expect a proof! Instead, I'll rehearse the (empirical) considerations which make
it likely that natural languages are inconsistent. The first point to make, however, is a methodological one. Consider the following from Charles Parsons
1984, 32–33, where he too tries to cast doubt on the suggestion that natural
languages can be inconsistent. He writes:

> In the specific case of the Liar paradox, the claim [that speakers of the language
> accept inconsistent inferences] would be that speakers are disposed to accept all
> inferences of the form: from '*p*' said by *S* on occasion *O*, to 'what *S* said on occasion
> *O* is true', and vice versa. That such a general disposition exists is highly likely.
> However, the empirical evidence (impressionistic to be sure ...) hardly bears out
> the idea that this amounts to a commitment to be honored come hell or high water.
> Confronted with the Liar paradox argument, almost anyone will recognize that
> something is fishy. Most would doubt that they fully understand what is being said
> and suspect nonsense. The difficulty there is in agreeing on a "solution" can be
> interpreted to mean that speakers do accept these inferences as general principles
> and do not know what to believe when they see the difficulties that arise in this
> particular case.

He continues this way:

> This view of the situation gives some scope for theory in dealing with the paradox. It
> may be that if we treat *all* the dispositions to accept statements and inferences
> involving 'say', 'mean', 'true', and 'false' as on the same level, the only conclusion we
> can arrive at is that it is impossible to attribute to all these words at once a coherent
> meaning. But a theory might be quite coherent and honor *most* of these disposi-
> tions. . . . Then the others can be attributed to confusion, difficulty of understanding
> expressions used in very abstract contexts, and the like.

---

[17] Herzberger (1967) offers an argument against the inconsistency view of natural languages
that doesn't rely on the notion of analyticity and instead focuses on truth conditions. This
argument is echoed in C. Parsons 1984: I'll consider it in 5.7. I should add that Herzberger
(1966) characterizes analytic sentences as true; although this follows venerable philosophical
tradition, it begs the question against his opponent: Either the principles presumed by natural
languages are inconsistent (and therefore, by definition, not analytic), or we define the principles
presumed by natural languages to be analytic, and then it's open to them (empirically speaking) to
be false. One can't *stipulate* that the principles presumed by a natural language *are true*. (And
one can't escape the inappropriateness of stipulating the properties of something—something
empirical—by describing the issues as not involving facts about language but (Herzberger 1966,
263) "more properly about the conceptual framework we employ in thinking about language.")

Parsons here tempts us with a very dangerous strategy. For, although he doesn't say so, to attribute *some* of the dispositions to confusion, difficulty in understanding expressions used in very abstract contexts, and the like, is to offer an empirical promissory note. One can't simply *claim* that speakers are confused, have difficulty understanding (certain) expressions in (certain) contexts, and the like; one has to *empirically verify* such claims. That's to say, it isn't up to the theory of truth, which negotiates a solution to the family of liar paradoxes, to draw the line between when genuine semantic dispositions are at work and when what's involved are what are often described as "performance factors"; that someone is confused, and so on, is a matter to be established empirically, and established in a way that's independent of the theory in question.[18]

This strategy is especially inappropriate for liars and strengthened liars because, contrary to what Parsons suggests, these sentences *aren't* hard to understand, and the unwelcome reasoning that leads to contradictions isn't hard to understand either. I quite agree that the ordinary person suspects there is something "fishy" going on; but it's hardly insignificant that whatever that person then manages to dredge up as the mistake that led to the contradiction either isn't very plausible—for example, that the statement is "nonsense"—or else (e.g., "neither true nor false") quickly yields a strengthened liar paradox.

The natural conclusion, then, *really is* that natural languages are inconsistent, just as Tarski thought, and for pretty straightforward reasons: Whatever principles are empirically discovered to be those that the proficient in the vernacular employ, they won't be ones that treat idioms such as "true" and "false" predicatively, nor ones that involve conditions on statementhood that imperceptibly go beyond what ordinary speakers already recognize as required for statementhood (e.g., presuppositions).[19] Given that this is right, certain natural worries arise. The first, and perhaps most pressing, is: Doesn't triviality ensue? Isn't *every* sentence (therefore) both true and false?

---

[18] I'm not, of course, claiming anything so grand as that the categorization of evidence for a theory must always be established independently of the theory in question—as everyone knows, that's a demand that's impossible to meet in general. I *am* claiming, however, that the particular psychological categories Parsons has helped himself to are ones that, in fact, are amenable both to fairly reliable folk claims about when they are and aren't operating, and ultimately, to fairly subtle psychological experiments of a familiar kind, e.g., measurements of response time to prescribed tasks, etc. The point, anyway, is that because notions of "confusion," or "difficulty understanding," are robust, psychologically speaking, they can't be treated as convenient dumpsites for whatever intuitions, or more broadly, dispositions, that fail to fit one's favorite semantic theory. Nothing in what Parsons has written here, by the way, indicates that he is unaware of the methodological point I'm stressing, except for his use of the optimistic word "scope." On the contrary, I would have thought that bringing in empirical factors of this sort *increases*—in the Popperian sense—the prior likelihood of the full theory (semantics plus psychological claims) being refuted by empirical testing.

[19] Again, I need to stress that I'm making an empirical claim; perhaps various psycholinguistic experiments will show that speakers *are* naturally confused when faced with (simple cases of) self-reference, and that certain principles which can be shown to generally apply everywhere else, are, because of this, misapplied in self-referential contexts. But I doubt it. We've been living with liar paradoxes for a very long time, and this, I think, is because, for just as long (at least), such paradoxes have been implications of the ordinary languages that people speak.

Yes. But I hasten to add that my empirical claim turns on whether the inference rules of natural languages are sufficiently like those of classical logic in this respect. Generally, ordinary speakers don't seem to think that every sentence follows from a contradiction, but (so I argue) this is because intuitions about relevance prevent people from recognizing that the very principles they do use allow them to draw every sentence as a conclusion from a contradiction.[20] Of course it's possible that I'm wrong, and that the principles involved in natural languages are ones that respect relevance (in some way).

5.5 Why the Inconsistency of Natural Languages Doesn't Make Them Useless

My claim, unlike the paraconsistent view of natural language, however, immediately raises the question of whether natural languages' presupposition of a trivial implication relation is *itself* a claim that's inconsistent with the undeniable fact that those proficient in natural languages have no problem reasoning within them, nor any problems with the making of ordinary truth-endorsements of various sorts. Many philosophers—perhaps most—believe that if ordinary languages did imply that every sentence is both true and false, then ordinary languages would be unusable.[21] I now show this is false.

Let's start by recalling the distinction between *implication* and *derivation*. Implication is a semantic notion, and since Tarski it's been (more or less) characterized this way: A set of sentences **A** implies a sentence B if and only if: If all the sentences in **A** are true in a model, then B is true in that model as well.[22] Derivation, on the other hand, is what's traditionally described as a syntactic notion: A sentence B is derivable from a set of sentences **A** if and

---

[20] Ordinary mathematics, that is, mathematics *as it's practiced in the vernacular*, draws much of its interest from the fact that proofs establish surprising results that no one expects. These results are unexpected precisely because ordinary mathematical proof crosses the barriers raised by intuitions about what's relevant to what. See Azzouni 2000a for further discussion of this. I should also add that one reason I'm largely convinced that one or another paraconsistent logic (one in which contradictions don't imply every sentence) *isn't* operating in ordinary language is because it seems to me that ordinary intuition licenses absorption, and given that, ordinary intuition can be brought to accept that a contradiction implies every sentence. See Priest 1987, pp. 103–4, for a discussion of this principle.

[21] Although Priest (1987) doesn't, as far as I can tell, straight-out *say* this, he clearly thinks it's a virtue of his dialetheism that the existence of sentences that are both true and false—the existence of dialetheias—*doesn't* imply that every sentence is a dialetheia; one indication, therefore, that Priest thinks of implication would make natural languages unusable is that, although he offers arguments for why we should think ordinary languages are inconsistent, he offers none to establish the equally important empirical claim that his dialetheism, or in fact, *any* paraconsistent logic (one in which inconsistency doesn't yield triviality) is the logic presupposed by ordinary languages. He also writes (1987, 104) after establishing the triviality of the assertion principle in the presence of a certain sort of self-referential sentence: "Thus semantics (or set theory) based on a logic which contains the assertion principle, is trivial: everything is provable. It is therefore suitable for no purposes, dialethetic *or otherwise*" (italics added).

[22] I will be taking a much closer look at this characterization of implication in part III.

only if there is a *proof* of B using (at most) as premises, apart from logical axioms, only sentences in **A**. To speak somewhat metaphorically, derivation is the operationalization of implication: Derivations are what speakers have access to; implications are what those derivations reveal.

As a result, although the definition of implication (syntactically speaking) focuses only on the sentences themselves, the definition of derivation also focuses on the stepwise derivational relationship—between sentences—that's mediated via other sentences. In consistent systems, this stepwise relationship is crucial: One can show that certain sentences can't be derived from other sentences *without* the use of certain intermediaries. Indeed, the need for intermediaries is a fact of mathematical life revealed by the ubiquitous use of the term "corollary."

Yet another way to put the above point is that derivation is intrinsically local: Sentences connect to other sentences via a network of fellow sentences. This is why Quine's old characterization of our body of beliefs as a network or web of beliefs is a metaphor with content: That content is the derivational roles that sentences must play vis-à-vis each other. And this is why Quine's locating of sentences within this network as more (or less) central also has content: Centrally located sentences play an ineliminable role in numerous derivations of implications from other sentences; peripherally located sentences don't.[23]

But this fact in turn leads to another: This is that contradictions, too, are local, even in a classical logical context where they imply every sentence. If the locale of a contradiction is central—say, a contradiction in a (oft-used) mathematical principle—that contradiction will impact globally on the web of beliefs. It truly does, in that case, endanger the coherence of the web of beliefs because so many derivations will have depended on it.[24] On the other hand, if the locale of a contradiction is at the periphery, it can be ignored, and then it won't infect the web of beliefs at all.[25]

---

[23] I've elsewhere (Azzouni 2000b) criticized this picture for overstating the derivational connections among the sentences we (collectively) take to be true. But this issue isn't germane to the current topic, and so, for the sake of expository ease, I'm (largely) leaving it aside—see note 30, however.

I should also add that *centrality* here is understood in terms of derivations that we (collectively) actually *draw*, and not in terms of the ones that, mathematically speaking, *exist*. In that sense, I'd imagine, no sentence is any more central than any other.

[24] I'm not thinking here of crude uses of it (when one is aware that it's a contradiction)—where, that is, every sentence is derived from it by a transparent use of its contradictory nature. Keeping in mind that contradictions needn't be easily recognizable, what can be derived—inadvertently—from such a contradiction is a complex tissue of inconsistencies. When the contradiction is discovered, and one wants to patch the result up, it won't be clear what to sort out and retain, or how. I'll make more of this point below.

[25] Recall that Lakatos once described all scientific theories as born refuted. More reasonably understood, this claim amounts to there being inconsistencies in our web of beliefs, more precisely, inconsistencies between certain statements (call them "observationals") taken to be implications of a theory and certain other statements that we've verified. This doesn't precipitate a "crisis" if these inconsistencies are localized in their impact on our web of beliefs.

These points, I hope, make clear why, if a contradiction infects a theory (or a language!), the scope of the infection will be overrated if one mistakenly focuses on implication—and the fact that such contradictions imply every sentence—instead of focusing on inferential practice, where it's clear that the scope of the infection can vary from dramatically destructive (if contradiction infects a central principle that's used regularly and widely) to (relative) insignificance (if it infects a peripheral statement that we're prone to overlook in any case).

The analysis I've just given still overstates how significant a contradiction can be, how much, that is, it can really render "incoherent" the results that have been drawn from it in a derivational system. For even if a contradiction infects a central plank of a theory, it needn't prove particularly damaging. A wonderful example of this may be found in the history of set theory: once informal set theory was replaced with one or another formal analogue (for most of them, ZFC), practitioners set about replicating the results of the informal theory, often using extremely similar proofs. This shows that the impact of the contradiction in the comprehension axiom was, in significant respects, minimal.[26]

Pretty much the same lesson may be drawn from the fact that liar paradoxes are implications of the general presuppositions of natural languages. These inconsistencies arise from two central planks: the naming and description conventions (i.e., that, roughly, we can call anything anything we want; and that we can successfully describe anything we can circumscribe in words) *and* the set of T-biconditionals. But, despite this, the contradictions so derived are severely localized in two respects. The first is that it's always specific sets of *truth-endorsing* sentences that are rendered contradictory, and these that don't have very wide scope—they aren't wide-ranging mathematical principles, for example, nor are they scientific laws.[27] Further, they *don't* affect other applications of the principles to more common (and valuable) uses of the truth predicate. This brings us to a second (related) point. Paradoxical truth attributions, and more broadly, sets of *circular* truth attributions, contrary to the impression that some philosophers have given, are seen by ordinary speakers as recherché.

---

[26] As I said, this needn't be the case; it all depends on whether what's been derived using a contradictory principle are themselves (as a group) largely consistent or not; there is nothing that can be said a priori about this—it depends on the case—that is, on what the derivational practice has *done* with the principle. This is why, although inconsistency is always threatening, it doesn't always make good on its threats.

I should add that my very brief description of the response of mathematicians and logicians to contradiction is very much potted history. For one thing, there is evidence that the profession—apart from Frege—decided only slowly that informal set theory was inconsistent; in this sense it doesn't seem that the comprehension axiom was seen as a principle of that set theory. However, the description, as it stands, seems to describes accurately Russell's response to his paradox. See Moore 1980 and Moore 1982 for nuanced discussions of the emergence and development of set theory.

[27] Notice the point. *This* infection is localized to cases where we use a *truth predicate*; it doesn't spread—more generally—to where we assert truths (or make denials)—since that *can be*, and often is, done without use of the truth predicate. The reader may fear I'm trivializing the issue: Sentences must have nontrivial *truth conditions* if speakers are to successfully use them—and this is so whether a truth predicate is explicitly used or not. *Not so*, as the earlier discussion in this section should have made clear. In any case, I discuss this concern in 5.7.

It's true, as Kripke (1984) has pointed out, that ordinary truth-endorsements can give rise to them easily; but it's also true that when speakers become aware of the contradictions that have emerged, they can reformulate what they "meant" to say since, except for occasional cases of willful philosophers, they certainly weren't trying for anything particularly self-referential in the first place.

As I've intimated above, one powerful test of whether an inconsistent derivational system is coherent or not is whether it can be, for the most part, embedded in a consistent derivational system which leaves intact most of the derivations employed in the original system. According to this test, our ordinary truth-endorsement practices are coherent since they can be (for the most part) embedded in a consistent predicative approach.[28]

One last point about this. Priest (1987, 5) writes:

> The natural presupposition [about the linguistic principles of our language] is that of inconsistency. For language and the principles that govern it have developed piecemeal and under no central direction. As logicians know, inconsistency is the natural outcome of spontaneity. Consistency has to be fought for. Therefore, *prima facie*, it would indeed be surprising if our concepts were internally and mutually consistent.

These are wise words, especially the ones about having to fight for consistency; nevertheless this gives the wrong picture of our language. Even if its principles are piecemeal sets designed to apply (at least initially) to specific domains of discourse, and ones that developed independently of each other,[29] the upshot usually is the apparent irrelevance of one part to another, rather than inconsistency between them. Often, what has to be fought for (or, more accurately, forged) are bridges of one sort or another, connecting discourse of one sort to another.[30] With respect to contradiction, what we have found in contrast to what Priest suggests, is that initially unrecognized inconsistency is localized to particular principles. Further, the inconsistency in question is often not even recognized until some particularly clever person exposes it. Since there is no mechanical test for inconsistency, this is to be expected.

5.6 The Norm of Consistency

As I've said, I think there are good reasons to deny that the logic presupposed by natural languages, despite their inconsistency, is paraconsistent. I've given

---

[28] As I argued in chapter 4, the kinds of circular but successful truth-endorsements that Kripke (1984) and Gupta (1984) describe aren't essential to our ordinary truth-endorsement practice; but the point being made here is a slightly different one: The coherence of our truth-endorsement practice is due to the fact that the vast majority of it is predicative, and that predicative endorsements are untainted by the contradictory nature of the principles of ordinary language that license truth-endorsements—and this is proved because those practices (for the most part) can be embedded in a consistent predicative approach.

[29] This, too, is an empirical claim, and one those who see language capacity as largely innate might deny.

[30] This is the lesson, anyway, that the special sciences seem to teach.

one reason already. Here is another: *It's an obvious and universal norm of ordinary speakers that contradictions are to be avoided.* Furthermore, *the* natural response to a paradox is to try to defuse it by ferreting out an assumption that can be rejected. Priest (1987, xv) acknowledges this when he writes, "Dialetheism *is* outrageous," but then he trivializes the point by immediately adding, "at least to the spirit of contemporary philosophy." No—it's outrageous to (pretty much) anyone proficient in the vernacular. The interesting thing is that ordinary speakers soon give up on such paradoxes (or, sadly, their obsession with paradoxes, not abating, drives them into philosophy); in doing so, it isn't that they presume that, after all, it might be that a liar paradox actually *is* both true and false: Instead they presume that something must be wrong, but they can't figure out what it is. (*"You* figure it out," someone once told me, a remark implicating both that something *is* wrong, and that, in any case—especially since life is short—the matter isn't very pressing.)

This point is worth stressing in a second paragraph. One of the marks that "inconsistency is *not* to be tolerated" is a norm of natural language is that speakers have no resources for handling paradoxes *other than* either finding (and rejecting) an assumption that the paradox depends on, or presuming (without being able to exhibit it) that, of course there *must be* a false assumption involved. Dialetheism (and paraconsistent logics, more generally) are recent innovations without genuine precedents in ordinary-language practices (Hegel notwithstanding).

This yields a very cute result: Although natural languages *are* inconsistent, ordinary speakers *don't believe* that they are. In fact, as far as I can tell, almost *no one* believes that ordinary languages are inconsistent. And this is hardly surprising since the avoidance of contradiction is, as I said, a norm of ordinary language; and *this* means that it's *rational* for ordinary speakers to believe that ordinary languages aren't inconsistent—despite the presence of evidence (e.g., liar paradoxes) to the contrary!

5.7 The Truth Conditions of the Sentences of Inconsistent Languages

I argued in 5.5 that inconsistency and its entailment (in non-paraconsistent contexts) of triviality don't in turn entail uselessness. But even those convinced of these claims may still be worried. Parsons, echoing Herzberger (1967), raises the issue of the cogency of assigning truth conditions (descriptions of circumstances under which sentences are true) to the sentences of English, given the assumption that it's an inconsistent language. He writes (1984, 32):

> An assignment of truth-conditions to sentences of English would have to satisfy some conditions of coherence, since it is not just an account of the beliefs of English speakers but an account of the conditions under which what they say is *in fact* true. But then it is hard to see how such an account could make English inconsistent.

The dialetheist, since he does not think inconsistency makes every sentence both true and false, has a response to this. What's mine? Well, I have two. The first—and this is a point about natural languages—is that it's simply true that if a language (or a theory) is inconsistent, then a theory of the truth conditions of its sentences entails that those truth conditions are trivial too: Every sentence is true (and false) in every circumstance![31] But it's also true that if the inferential practices of the users of a language are coherent, then even if the implicit principles of the language trivially imply that every sentence is true, users of the language won't act that way. And so, we can capture the coherence of their natural-language practices, not by giving truth conditions for the sentences of those languages, but by supplying a (consistent) regimentation of their practice in that language. That is, we can engineer a (piece of) artificial language that's consistent[32] and capture in the sentences of that language (and the inferential relations among them) a portion of ordinary language (that we want) with *its* sentential vehicles, and the inferential relations empirically established to hold between those ordinary-language vehicles. In doing so, of course, we can leave out certain awkward constructions of ordinary language, if they don't serve a useful role (and, in this case, if they are the—or one—source of the inconsistency of natural language); and, apart from this, we can also engineer the inferential relations of the artificial language in ways that are more convenient or efficient or otherwise appealing than the way they appear in ordinary language. In this sense regimentation is a *normative* activity.

Because, near as we can tell, the regimentation is consistent, we can also use it to stipulate truth conditions for natural-language sentences by the truth conditions we give to the proxies of those sentences in our engineered language. These truth conditions, unlike the ones directly supplied to an inconsistent language, won't make every sentence true under every circumstance. What licenses our attributing these truth conditions to *natural-language* sentences, despite the fact that, strictly speaking *their* sentences belong to a different (and inconsistent) language, is that the truth conditions we supply to their proxies capture—nearly enough—the inferential practices (and truth-endorsements) that natural-language users engage in. The semantics we give to the artificial language, that is, matches the inferences mathematically licensed by, say, completeness and consistency proofs, and these in turn mimic (nearly enough) the inferences that natural-language users engage in with natural-language sentences.

---

[31] On some views of how truth conditions must be given—via a translation into the language of the theory of those truth conditions—the attribution of such truth conditions to the sentences of any language will infect the language of the theory of truth conditions so that such a theory must either collapse into triviality as well, or *deny* the inconsistency of the language so translated. Two points about this: First, I don't believe that the giving of truth conditions requires translation, so the feared spread of trivialization is unfounded. I intend to discuss this claim in later work. Apart from this, the general approach of regimentation that I've been arguing for in chapter 4 deserts the project of directly supplying truth conditions for inconsistent (natural) languages, and instead supplies truth conditions for the (consistent) regimentation. See what's forthcoming.

[32] Qualification: As with any piece of mathematics, we can never be absolutely sure that a regimentation *is* consistent. (There is no mechanical test for consistency.) But we can be pretty sure that it's consistent, and that's the best we can ever do.

We take liberties (this is our *right*) with the inferences of a natural language we regiment: We ignore some, capture others with modifications, perhaps coin new ones; we take further liberties in denying some sentences of the natural language any representation at all. This is what—in chapter 4—I suggested we should do with the self-referential sentences that accidentally arise in the vernacular: Our regimented language—at least as far as truth-endorsement is concerned—should be chosen to be one with a hierarchy of AU-quantifiers, each one with a domain strictly inclusive of the ones below it. There is no place in such a hierarchy for circular truth-endorsements (there is no problem representing, of course, circular reference apart from truth-endorsement).

Precisely here dialetheists, who have been chased from the Eden of natural language, have a second chance. For even if they agree with what I've claimed about natural languages, they can argue that the appropriate choice for the regimentation of natural languages is an artificial language based on a paraconsistent logic. This raises issues that I must put off for another time. Perhaps the most substantial is this: The dialetheist will claim that such regimented languages can be designed to be semantically complete in the sense given earlier: Anything that can be said about them, including laws governing their semantics and syntax, can be said in them. This isn't true of the predicative languages I'm offering. Also, there will be statements, ones, for example, that describe the predicative hierarchy as a whole, that will be inexpressible in the predicative approach; such inexpressibles needn't exist in paraconsistent constructions. Does this definitively advantage paraconsistent regimentations over their classical (and predicative) competitors? I don't think so, although my claim is not obvious, nor easy to show.[33]

---

[33] And so I won't be trying to show it in *this* book.

CONCLUSION TO PART I

Most truth deflationists focus on the T-biconditionals, and read their deflationism off of a kind of redundancy intuition—that to say "*S* is true," for any sentence *S*, is to say something equivalent (in some sense) to *S* itself. Using this humble beginning as a basis, some try to claim that (therefore) truth isn't a property; others say that there are several notions of truth—all richer than the minimal redundancy one.[1]

The objections and troubles that this species of deflationism faces are due to this very same assumption; for it simply isn't powerful enough to yield what such philosophers have claimed on its behalf.

I've taken a different route. I've accepted the Quinean dicta that the functional role of the truth predicate is semantic ascent and descent—when coupled with an appropriate quantifier. Strictly speaking, this does *not* yield deflationary conclusions about *truths*—that nothing in general can be said about, e.g., the correspondence role of true sentences. I believe this latter claim, but I also believe it must be established on substantial metaphysical grounds, not on slender logical grounds solely about the role of the truth predicate, or whatever idiom one adopts instead to manage that role.

I want to rehearse, briefly, the claims I've argued for—in other work—on behalf of my metaphysical truth deflationism. First, I've claimed that semantics in general, and Tarskian semantics in particular, don't themselves require any ontic commitments, and that this is so even though Tarskian semantics involve "domains." For those domains are themselves characterized by

---

[1] For the second option, see Wright 1992, Lynch 2000.

descriptions (and thus quantifiers) in the metalanguage, and so we commit ourselves to the existence of objects in such domains only insofar as the quantifiers we use there do so.[2]

I thus replace the recognition of ontology by the syntax (regimented or otherwise) of a language we speak, with the substantial inspection of ontic predilections—in ordinary life and in the sciences. I claim that such an inspection yields a denial of commitment to various classes of objects—fictional items, mathematical abstracta, and so on—regardless of whether quantification over such items is indispensable or not.

The position described so far is compatible with a Wright-style one that treats different discourses as utilizing different truth predicates—some deflationist, some not. So, on such a view, fictional and mathematical discourse would be characterized by a deflationist truth predicate, but discourse about physical objects would involve a more substantial truth predicate.

I believe this Carnapian-flavored view is ruled out by the way that the sentences of our discourse are bound up with one another. The tightness of the bonds isn't easily seen if one considers, for example, applications of counting to zebras, where a mathematics-free language is available and already in place for describing that kind of object. One has to look at examples in the sciences where the language of the empirical theory is constituted all-at-once (as it were) from mathematics. Although programs of distillation (of sentential empirical subject matter out from the pure mathematics that's applied to it) have been attempted in such cases, they haven't succeeded, and there are very good reasons to think they can't succeed.

As a result, I claim that one can't determine one's ontic commitments, and therefore, the cases where one's sentences are to be understood as describing a correspondent reality, by looking at sentences on a case-by-case basis. Instead, one must look at the terminology used *in* such sentences—and see in which cases the terms robustly refer to objects external to ourselves, and in which cases the terms "refer" only in a deflationist sense—that is, to nothing at all. But if this is right, then truth—applying as it does to *whole* sentential vehicles—just isn't a tool for *ontology* because truths don't sort out neatly into different kinds; and so only *one* truth predicate—if we opt for *a* truth predicate (at all)—exists.

I should take up one other family of issues before moving on to part II.[3] This is the question of exactly what the truth predicate in English is doing. A first stab at the family of issues: What does it mean to say "S is true"? To some extent, this is a bad question to ask; it's not the question: What set of sentences is picked out by the truth predicate? but rather, What am I saying about them when I claim they are true? Compare this with the predicate "begins with the letter 'r'." The sentences, all of which begin with the letter 'r', have numerous properties in common (let's say); but when I attribute that pred-

[2] See Azzouni 2004a, 53–55.
[3] My thanks to Douglas Patterson for questions that prompted the forthcoming line of argument.

icate to them, I'm focused on the particular property *begins with the letter 'r'*. What's (similarly) going on when I attribute truth to a sentence?

It's unwise to press the "begins with the letter 'r' " model for predicates too hard—predicates are of all types, and we don't always have something "intensional" in mind when we use a predicate; we don't need to do so for the predicate to operate successfully.

Well, okay, someone might respond to me, but (still) how does "true" in English work? It's not (really) an anaphorically unrestricted quantifier— you've admitted that's a regimentation. And, given your opposition to treating T-biconditionals as analytic in the sense that the right side has the same content as the left side, "true" doesn't operate by transferring content anaphorically—as prosententialists might take it to operate. So how is the trick supposed to go?

Well, start by taking the T-biconditionals to fix the reference of the truth predicate in such a way that blind endorsements are appropriately fixed in truth value. *But wait*, someone is sure to protest, these things are just a list, and you've denied that they're a complete list. In chapter 2, you argued that such lists won't fix blind endorsements of sentences in foreign languages; similar arguments show they won't fix blind endorsements for sentences of our own (future) language, when new terminology is introduced that's not translatable to ours. It's also likely—because you've ruled out propositions—that they won't fix current but demonstrative-laden sentences of our own language.[4] Furthermore, there is no easy way—in the vernacular—to determine when a T-biconditional is inconsistent or not. So this (on your view) isn't going to work.

Well—that's right—*it won't*. But all these indicators that a list of T-biconditionals of the sentences of a language can't do the job *don't* mean that such T-biconditionals *aren't* the (sole) resources that speakers have for fixing the extension of the truth predicate. Let me explain by switching the example to another that I've discussed elsewhere.[5] Consider what are called "natural kind" predicates. My view is that such predicates don't, in any robust sense, pick out their (entire) extensions. Rather, we talk as if they do, while the real work of picking things out, of connecting to the world, is done by the causal relations we forge and refine to items in the world—relations that change over time. I call these "procedures," and the idea is that the set of procedures— over time—is not only augmented but also diminished: We can decide that certain ways we had of getting to something didn't really get to it at all.

On this view, we talk as if the reference of "lead," for example, is an un-changing extension—but our practice is one of developing ever more subtle and refined ways of causally identifying lead. These causal ways are treated in our *talk*, not as a metaphysically grounded referential grip on lead, but as epistemic links to something that we *already* successfully refer to. Those who

---

[4] Even though speakers may be willing to supply demonstrative-free statements that "come to the same thing" for *some* demonstrative-laden sentences, it's highly unlikely they could (or would) do that for all demonstrative-laden sentences—recall 2.5.

[5] In Azzouni 2000b, especially part IV.

want to cement reference causally[6] need to invoke some sort of view that involves "causal covariation"—something like a view that under appropriate conditions competent speakers tend to utter a predicate "P" when and only when they are causally impacted by instances of P-ness.[7] There are reasons, however, to think such a view is hopeless—that even its apparent plausibility relies on the illusion that the ways that we're causally impacted by things of a kind happens in a relatively homogeneous way. In fact, there are always an unlimited number of exceptions—our ability to utter "P" when and only when causally impacted by instances of P-ness is an ongoing and open-ended practice of refinement, one that can't be characterized in a neat way ahead of time.

In this respect, the truth predicate is no different—nor should it be. We speak—that is, we blindly truth-endorse (and blindly false-endorse) sentences howsoever we need to, and regardless of whether our own language—and its T-biconditionals—successfully enables that endorsement: We talk as if every sentence (or proposition) whatsoever falls under our truth predicate or under its anti-extension. In practice, however, we forge access to sentences somewhat similarly to how we forge our way to instances of lead. In the latter case, it's a matter of developing more subtle causal relations to instances of lead located in hard to reach places. In the case of sentences, it's a matter of enriching the vocabulary of our own language so that sentences heretofore inexpressible in our language subsequently become expressible.

The problem is this: Although we can recognize the role of the truth predicate as for a particular job (semantic ascent and descent in the context of blind truth-endorsements)—the resources, *within the vernacular*, for successfully semantically fixing the role of that predicate are solely T-biconditionals. This makes it easy for someone—like Horwich—to think that the T-biconditional scheme is the whole story about truth because nothing else can be expressed in the vernacular that governs the truth predicate. That is, the expression of a suitable generalization, of the sort available via AU-quantifiers, isn't available in the vernacular—and this despite our sensing that the meaning of the truth predicate transcends T-biconditionals (we sense its meaning as linked to a generalization that we nevertheless can't express in the vernacular: a generalization directly linked in turn to the function of the truth predicate).

Despite the above, as I said, we treat the reach of our truth predicate as the whole field of what is—in principle—truth-apt, even though our actual resources fall woefully short. In practice, we're always refining the list of T-biconditionals: winnowing items that breed contradiction (or ignoring them) and including new T-biconditionals for new vocabulary.

What about what the truth predicate *means*? Well, its "meaning" can't be given by the list of T-biconditionals at a time—that list is always changing, and

---

[6] And really, what other option is there? Some might suggest properties—eligible classes in Lewis's sense. But this will achieve by sheer verbal fiat what we need done *in reality*. (See Azzouni 2000b, part III, esp. 192–95.)

[7] This kind of view is fairly popular. See Stampe 1977, Dretske 1981, and Fodor 1990, among many others.

so too would the meaning of the word "true" change along with it. (We would have to recognize ourselves as operating with a new truth predicate every time we mint new vocabulary.) Instead we treat the word as the same one, and one with a meaning that somehow gives us the list to T-biconditionals.

But there are no resources in *the vernacular* to make good on this promissory note about meaning *apart from* our recognition that the role of "true" is to facilitate blind endorsement. To express this in the vernacular requires a device of generality: And I've argued in part I that only invented terminology, AU-quantifiers, can do that job.[8]

Here's an objection to the above: You're telling a story about how the truth predicate operates in ordinary discourse that—semantically speaking— you admit *won't work*. It will—strictly speaking—give the wrong answers in cases of blind endorsements to, e.g., foreign truth vehicles not translatable to our language, or (arguably) even in cases of demonstrative-laden sentences already in our language. That means it *can't be* the right story about how the truth predicate operates in the vernacular. Answer: Yes it can—provided we can explain why (almost all) speakers will regard it as giving the right answers. Here the lessons of chapter 5 should be taken to heart. It's a much more dramatic result that natural languages are inconsistent than it is that the semantic resources for assignments of truth values to blind endorsements— strictly speaking—give answers at variance to what our practice with those endorsements requires. And yet, (i) it can be explained why speakers think (falsely) what they think, (ii) it can be explained why there is no pressure on the practice in the vernacular to recognize the inconsistency, and to devise a strategy that resolves it,[9] and in any case, (iii) the practice of the vernacular can be regimented into something more in accord with the implicit assumptions of that practice. Exactly the same is true of the blind-endorsement practices in the vernacular.

If the only resource that natural languages have for fixing the extension of the truth predicate is the T-biconditional, then the working assumption of the transcendence of that truth predicate is flatly inconsistent with those resources. Here, as I've argued, is where AU-quantifiers come into play: Their existence shows that a regimentation exists—a consistent construal of our blind-endorsement practices. Regimentations, thus conceived, are quite similar to mechanically recognizable derivations. I'll claim (in part II) that mathematical proofs indicate mechanically recognizable derivations—even if these derivations are impossible to exhibit explicitly. In the same way, our practices in the

---

[8] I made this point in 3.5, but it's worth stressing yet again: AU-quantification over sentences involves anaphora (of course) but the anaphora involved doesn't require shifting content from the items quantified over to the places where the variables occur.

[9] Here there is an evolutionary analogy with the more unpleasant aspects of the phenomenon of aging: There isn't (much) selective pressure to remove the increasing declines that age generously bestows upon us, as long as that bounty emerges after the reproductive phase is over. So too, if our truth-endorsement practices fall afoul of examples that can either be easily ignored, or can be—after the fact—treated as if they fell under the rubric of the practice all along, then (ordinary) speakers won't feel the semantic inadequacy of their practice.

vernacular—that we can't desert—are ones that we, via what we show about regimentations, know to be reformulatable in consistent ways. We can continue to endorse the truth or falsity of sentences in an inconsistent medium because we know that what we do—for the most part—can be preserved in a consistent regimentation.[10]

In light of the fact that I'm offering a description of a semantic practice for practitioners of ordinary languages that—strictly speaking—provides truth values for their blind endorsements at variance with the truth values they take those endorsements to have, some may want to revisit those places in the foregoing argument (e.g., in 2.7) where a substantial property, by virtue of which the truth predicate picks out what it needs to pick out, was rejected. Those so tempted should think again. This is a case where a broader perspective that considers other pseudo-solutions in philosophy may prove consoling. One is *often* tempted to postulate *properties* to facilitate explanations for how we grasp what we seem to grasp.[11] Doing this, often, is only trying to force a *label* (of our own making) to pick out what we need picked out without our offering any explanation for how the label is supposed to manage this. Just this maneuver is being engaged in by proponents of propositions who take us to grasp such things and their properties by sheer virtue of having *terms* in our language that *require* (on their take of the semantics of such terms) our grasping such things and their properties.[12]

Concern with the truth idiom is now going to be set aside for a while. Part II is concerned with the nature of mathematical proof. In part III, I return to truth—at least as it shows up in our (ordinary) notion of implication: The concern is to balance the results of part II against that notion of consequence, one that seems to involve truth.

---

[10] This excuse for what we do in the vernacular—in terms of its successful implementation in a regimentation—is sheer (intellectual) idealization. Most of us just continue our practice in the vernacular without any worries whatsoever. Only those of us whose consciousnesses has been raised—by the considerations of part I (let's say)—need to work through this rationale.

[11] One is often so tempted in philosophy of mathematics and philosophy of language—for example.

[12] Notice that the objection is not to properties per se—it's an objection to the sloppy epistemology, as it were, that so often accompanies such properties: postulated enablings by which our terms automatically refer to such properties.

II

MATHEMATICAL PROOF

Mathematics is not to be characterized by its objects (such as: space and time, forms of inner intuition, theories of number and measurement, and the like) but only, if one wishes to circumscribe it completely, by its peculiar process: the proof. Mathematics is a systematization of the provable and as such, an applied logic; its task is the systematic development of "logical systems", whereas "pure logic" only investigates the general theory of logical systems. Now what does "prove" mean? A "proof" is the derivation of a new proposition from other previously given propositions, by whose truth its own is established through general logical rules or laws.

Ernest Zermelo (cited in Moore 1980, 120–21)

INTRODUCTION TO PART II

The next two-thirds of this book take up, respectively, *proof and consequence.* Here I turn to the absolutely central question of—to put it grandly—the nature of Human Reason, and the sort of access we have to that Reason. My concern isn't with the ancient contrast between our ability to recognize what's reasonable despite our tendency to behave "irrationally." Rather, I'm interested in what sort of grasp we have of the reasoning that enables us to execute proofs—items the success (or failure) of which we can pretty much uniformly agree on.[1] In a way, this question goes straight back to the dawn of philosophy, to Plato's *Meno,* anyway. I claim—and although this will sound a little murky now, I hope to make it quite clear in the course of what's coming—that we don't have introspective access to how we reason, nor even to the principles, if any, *by which* we reason. This has dramatic consequences: It explains how what has long been taken to be a straightforward intuitive notion of semantic entailment: *if A is true, then B must be true,* is a notion that doesn't actually illuminate our grasp of what follows from what, but largely encapsulates our ignorance of how inference operates. As I said, I hope the subsequent discussion in this book makes clear what I am (and am not) claiming here.

---

[1] Marcus du Sautoy (in Longstaff 2003, 50) is quoted as saying: "In mathematics, when somebody has proved something, all the evidence is on the table. You've got a proof. OK, there'll be a bit of controversy when someone will say they have proved something, but has made a mistake and that is pointed out." And he illustrates this with the following anecdote: Louis de Branges's "proof got thrown out because it was rather incoherent. But after a working seminar in Russia, mathematicians there said: 'No, we're sorry, this is proof and he's done it'. And the mathematical community had to swallow its words."

In part II (specifically), I look at the paradigmatic example of successful reasoning—and concept formation—at its clearest: mathematical proof.[2] I aim to provide a theory of what exactly is being shown when a mathematical proof is given, and what sorts of resources mathematicians have for proofs. I see, broadly speaking, two schools of thought about this. The first—I'll call it "the establishment"—holds (roughly) that (one or another) formal logic, and the mechanically recognizable derivations that can be constructed on the basis of that logic, normatively govern such mathematical reasoning. The default position for establishment thinkers is that, in some sense, mathematical proofs, although almost never derivations in this technical sense, get their rationale from the mathematician's access to such derivations: More strongly, ordinary mathematical proofs are "abbreviations" of derivations. Those thinking of mathematical proof this way see the unique epistemic qualities of mathematical knowledge as pretty straightforwardly resulting from (implicit) reliance on such derivations.

The second school of thought—self-styled "mavericks"—presume it's manifestly clear that such derivations can't be the backbone of mathematical proof because mathematicians don't have access to, and are not trained to be conversant with, such derivations. Thus, philosophers, sociologists, and so on who are mavericks look to external sociological factors to force proof (at a time) to take the form it takes. It's no surprise that mavericks often invoke the rule-following problem attributed to Wittgenstein to motivate their view of mathematical proof.

As usual, when smart people are opposed and furthermore, almost amazed that others have different views, there is a great deal to be sensitive to with respect to both positions. I think a view is possible that accounts for the evidence that both schools of thought are (respectively) concerned with, and I provide one here. The broader concerns of the book emerge this way: The general view of proof I endorse apparently hasn't a substantial place for "truth" in it. Furthermore, the principles by which proof works (according to me) are so inaccessible to (even sophisticated) introspection that mathematics could flourish for more than 2,000 years—be one of the most successful disciplines we've ever had—and yet our understanding of what's really behind mathematical proof remain undisclosed until sometime in the *twentieth century*. This brings us to the topic of part III, where I turn to the conceptual notions we possess of implication—the semantics, as it were, of proof.

Chapter 6, the first chapter of part II, takes on the mavericks: It shows that the sociological forms of coercion that the uniformity of behavior in most areas depends on, including language, can't be at work in mathematics. In this sense mathematics is sociologically unique. This poses a challenge to mavericks because the external forms of coercion—the only ones they officially allow themselves access to—don't explain uniformities in mathematical practice. In turn, chapters 7 and 8 take on the establishment and show how mechanically

---

[2] It's no surprise that mathematical proof is widely taken as the paradigm of genuine knowledge as well.

recognizable derivations can't psychologically underpin how mathematicians see what follows from what. This sets up the dilemma about mathematical practice that motivates my solution—also presented in chapters 7 and 8: the derivation-indicator view of mathematical practice. Briefly, this is the view that a mathematical proof of B from A *indicates* that there is a mechanically recognizable derivation from (a proxy of) A to (a proxy of) B in an algorithmic system. Roughly speaking (in classical mathematics) the algorithmic systems are based on a logic rather like that of the first-order calculus—there is an effective notion of proof (in any case), and such systems use (again, roughly) the concepts, tacit and explicit, that appear in the ordinary mathematical proof. Apart from this, there is quite a bit of latitude (that doesn't matter) about exactly which derivation in which algorithmic system is so indicated.

The claim isn't that mathematicians recognize that this is what an ordinary mathematical proof does (indeed, the suggestion is that mathematicians got it wrong—when they theorized at all about mathematical proof—for most of the existence of the discipline). Rather, it's that the practice of theorem-proving in mathematics was shown (empirically) to be of this form by the success of regimenting mathematical proof in algorithmic systems based on first-order logic. Mathematicians recognized, and still recognize, the success of a mathematical proof by what (phenomenally) feels like a grip on a *semantic* relation—specifically implication—between the steps in that proof: a semantic relation that intuitively feels based upon what the statements in the proof say (what they're *about*). Here—as so often—phenomenal feels are utterly misleading. My claim is this: The elements of a traditional proof enable it to be correlated to a formal derivation, where all the aspects of the traditional proof that seem to rely on the properties of objects referred to have been replaced by explicit axioms. Thus the promissory note that a semantic relation always seems to offer (reference to an object) is here discharged by "disinterpretation"— that relation dissolving into explicit axioms sufficient for the construction of a mechanically recognizable derivation.[3]

In the previous course of this book, although I've more than once alluded to my nominalism (especially in 1.8 where I gave reasons for being an MTDist), it hasn't played a very substantial role. This doesn't change now; although it may look like that very nominalism is providing motivation for chapter 8. A case can be made (and is made in chapter 8) that mechanically recognizable derivations, the ones indicated by ordinary mathematical proof (according to the derivation-indicator theory) aren't instantiated in the world in a way that bears on mathematical practice. That is, such derivations can't be, for example, psychologically represented in the minds of mathematicians; they can't be the deep-structure correspondents of ordinary mathematical proofs. If right, these are the sort of considerations that mavericks have in

---

[3] I'm speaking here of the conceptual proofs so central to twentieth-century mathematics. Algebraic and numerical computations already—phenomenally speaking—feel quite syntactic: an apparent matter of manipulating formulas by rules.

mind when they deny that formal systems can be epistemically relevant to ordinary mathematical proof. Notice the point: The nominalist quickly worries about the status of such derivations because it seems that if no instances of them are in the world, then their effect on mathematical proof must be one of the mathematician (magically) grasping abstracta, and (according to the nominalist) there are no such things (even to magically grasp). But there is a problem even if we accept the existence of such objects. For the question still remains of how they play a role in mathematical proof: How does the ordinary mathematician, when constructing ordinary proofs, practice so that mechanically recognizable derivations are indicated?

Let me provide a solution to show what the problem is. Imagine (contrary to the arguments presented in chapters 7 and 8) that ordinary mathematical proofs are (strict) abbreviations of mechanically recognizable derivations. One would establish this, presumably, by showing that the collection of informal practices that mathematicians use to write down proofs preserves the strict abbreviation relation to (one or another) mechanically recognizable derivation. That is, the procedure would be very similar to what's done in Kleene 1971 when a derived rule for proofs is introduced. One shows that any proof in which applications of the derived rule appear *can be rewritten* so that only applications of original rules are used. Unless one invokes at crucial stages a literal perception (or psychological internalization) of the mechanically recognizable derivation that's to be indicated by an ordinary mathematical proof, one needs to at least point in the direction of how it is that the collection of open-ended concepts and tools introduced into mathematical proofs preserves the indication relation to derivations.

My analogous position, as I've stated, is that traditional mathematical proofs correlate to mechanically recognizable derivations. Here, too, to ultimately make the case for this, one *exhibits* a mechanically recognizable derivation correlated to a traditional proof. And, admittedly, I don't do *this*. But—as with Church's thesis—one points to empirical evidence, for example, the success of Russell and Whitehead's *Principia*.

One might worry that subsequent (traditional) mathematical proofs could falsify the derivation-indicator thesis: After all, the devices used in mathematical proofs are too open-ended (as just said), and topic specific: They quite often turn on the perceived properties of the objects mathematicians take themselves to be proving results about. But it's clear, nevertheless, how this is compatible with the derivation-indicator view. For, first, object-centered thinking is a matter of encoding axioms. When someone thinks of the counting numbers, he or she can write down (or *presuppose*) various facts taken to hold of them. In this way, the principle of induction (say) can come to mind, and this principle can prove independent of (and yet consistent with) the other principles (implicitly) used up until that point. What powers the picture of numbers that the principle was drawn from, of course, can be concrete imagery of the application of numbers in specific cases. But this imagery can be much richer and subtler, and can be driven by internal factors that only obscurely connect to the applications just mentioned. Its presence doesn't mean, and doesn't require, that the mathematician is grasping actual objects, and abstracting principles from *those*.[4]

Second, the derivation-indicator view never traps the mathematician within a single algorithmic system. The mathematician shifts from algorithmic systems to stronger ones almost unconsciously. What's required is that, when the mathematician proves new results on the basis of new principles seen, a derivation of the right sort exists. But this is satisfied by the mathematician having learned a practice that corresponds to a process of mechanical recognizability— even if that mathematician doesn't realize that such is what the practice comes to and that such is what's being indicated when he or she regards a proof as successful.

As I said earlier, that the derivation-indicator view of mathematical proof is right is something to be established empirically in *exactly* the way that, say, Church's thesis is established empirically: It can't be shown by a sophisticated examination of what the mathematician experiences when establishing proofs. The view also doesn't require the actual *existence* of mechanically recognizable derivations; all that's required is that the mathematician's practice be in accord—as it were—with a theory of such things; that is, that his or her ability to recognize phenomenally when *B* follows from *A* accords with (a formal proxy of) *B* being derivable from (a formal proxy of) *A* in the context of some appropriate algorithmic system.

But this raises a challenge. The derivation-indicator view is that ordinary proofs correlate to mechanically recognizable derivations, and it is because of such correlations that traditional proofs succeed. But the ordinary mathematician is unaware of such derivations, and the ordinary mathematician's recognition that a proof *does* succeed therefore doesn't turn on sensing the existence of such derivations (since, in fact, most such derivations don't exist). This seems to leave two options. (i) Either whatever it is that the mathematician does sense has semantic content—and therefore, on this view, the traditional mathematician recognizes successful proof via a systematic phenomenological error: The purported object the mathematician is thinking of, and the representation of such, is (actually) code for propositional (axiomatic) content that the mathematician hasn't phenomenological access to. Or (ii), on the other hand, the mathematician must (tacitly) sense that a traditional proof can be expanded into a mechanically recognizable derivation.[5]

---

[4] It doesn't have to mean this in the mathematician's case any more than it has to mean this in the novelist's case, where his or her depiction of a character develops richly, but yet in a way that still seems "true to life"; although neither the novelist nor anyone else can think of actual human beings like that character. Another kind of case are artistic depictions of "landscapes" that look like nothing we have ever seen (and yet are still recognizably "landscapes"). Such imaginative feats don't require a theory of a psychological pipeline to the "real things" on other planets.

[5] Actually, I think both these options are right, as the remainder of this book will illustrate. The mathematician often recognizes as immediate steps in inference the movement from the general to the particular, or modus ponens, both of which *also* exemplify mechanical recognizability. *And* the mathematician also thinks objects are being thought about because of the misleading phenomenology of proof—this is the main intuitive source for platonistic doctrines.

But why adopt either of these (unappealing) views? Why not take what it is that the mathematician recognizes to hold between statements A and B, when B is proved on the basis of A, only to be that A *implies* B? Why not claim that it's this implication relation that mathematicians (and everyone else, for that matter) recognize when they realize that B follows from A? Why not say that what the mathematician does is simply break that relationship up into subsidiary implication relations between successive items in a series of statements that can be immediately seen to hold? In the first-order case, of course, the implication relation (in question) is coextensive with the syntactic deductive relation. But in higher-order logics, this is not so, and our grasp of what's going on is thus clearly semantic: We *understand* the implication relation even though it isn't effective. I take up this challenge explicitly in part III.[6] One thing I *do* claim is that the mathematician *does* sense that any traditional proof can be expanded so that "no steps are missing." After doing so, of course, the mathematical object drops out of the picture since its role is replaced by explicit assumptions presupposed in the proof. The mathematician engaged in traditional theorem proving was unaware of this, and one reason for this unawareness is that the phenomenological sense of a guarantee—that arises when contemplating the steps of a traditional proof—isn't due to one's feel for the shape of a mechanical derivation that can be correlated to a traditional proof but to the psychological capacity on the part of the mathematician to bundle assumptions together so that they can be tacitly assumed without explicit awareness of them. I don't discuss this aspect of the cognition of traditional proofs any further in this book. (I developed the material too late to include here.) The interested reader should look at Azzouni 2005.

Part II is drawn, in part, from the three papers listed below. (1) nearly continuously appears as chapter 6, (2) nearly continuously appears as chapter 7, and (3) nearly continuously appears as chapter 8. In those papers, I thanked various people for help—I remain grateful for that help.

(1) How and why mathematics is unique as a social practice. In *Perspectives on mathematical practices. Proceedings of the Brussels PMP2002 Conference*, ed. J. P. Van Bendegem. Dordrecht: Kluwer, forthcoming.

(2) The derivation-indicator view of mathematical practice. *Philosophia Mathematica* (3) 12: 81–105.

(3) How to nominalize formalism. *Philosophia Mathematica*. (3) 13: 135–59.

---

[6] Although I must add now that the view currently under consideration is wildly implausible on the face of it: implication relations, in general, are quite unusable: On the model-theoretic view, for example, it's impossible to see how anyone can determine directly (i.e., by checking each such model) that in every model that A is true in, B is true as well.

6

The Uniqueness of
Mathematics as a
Social Practice

6.1 The Symptoms of Social Conditioning

I'm sympathetic to *many things* those who self-style themselves "mavericks" say about how mathematics is a *social* practice. I'll start with the uncontroversial point that mathematicians usually reassure themselves about their results by showing colleagues what they've done. But *many* activities are similarly (epistemically) social: Politicians ratify commonly held beliefs and behavior; so do religious cultists, bank tellers, empirical scientists, and prisoners.

Sociologists, typically, study methods of attaining *consensus* or *conformity*[1] since groups act in *concert*. And (after all) ironing out mathematical "mistakes" is suppressing *deviant behavior*. One way to find genuine examples of socially induced consensus is to limn the range of behaviors *possible* for such groups. One *empirically* studies, that is, how groups deviate from one another in their (group) practices. Consider admissible eating behavior. The options that *exist* are virtually *unimaginable*: in *what's* eaten, *how* it's eaten (in what order, with what *tools*, over how much time), how it's cooked, *if* it's cooked, what's allowed to be said (or not) during a meal, and so on. To understand why a group (at a time) eats meals as it does, and why its members find variants inappropriate

---

[1] Attaining "conformity" and "consensus" are mild-sounding phrases for what's often a pretty brutal process. Although what I say is intended to be understood generally, the reader does best *not* to think of the practice of torturing political deviants (in order to bring them into line), but of doting parents teaching children to count, to hold forks, to maneuver about in clothing, or to speak.

(even *revolting*), we see how consensus is determined by childhood training, how ideology crushes variations by making them *unimaginable* or viscerally *repulsive* (so that, say, when someone so trained imagines an otherwise innocent *cheeseburger*, what's felt is nearly instinctive disgust), and so on. Equally coercive social features in conjunction with those just mentioned explain why we obey laws, respect property, and so on: the threat of punishment, corporeal or financial.

Before turning to mathematical practice, note two presuppositions of any empirical study of the social inducing of consensus (and that are assumed in my sketchy delineation of the sociology of eating). First, such *social* inducing presupposes (empirical) evidence of the *possibility* of alternative behaviors. The best way (although not the only way) to verify that a kind of behavior *is* possible is to find a group engaged in it; but, in any case, if a behavior is biologically or psychologically impossible, or if the resources available to a group prevent it, we don't need social restraints to explain why individuals *uniformly* avoid that behavior.

The second presupposition is that the study of social mechanisms should uncover factors powerful enough to exclude (in a given population) the alternatives we otherwise know are possible. Either the absence of such factors, or the presence of empirical reasons that show such factors can't *enforce* behavioral consensus, motivate the hypothesis of *internal factors*—psychological, physiological, or both—in conforming individuals. Consider, for example, the Chomskian argument that internal *dispositions* in humans strongly constrain the general form of the rules for natural languages.

6.2 Conformities in Mathematical Practice

There are two striking ways mathematical practice differs from just about *any* other group practice that humans engage in. One of these striking ways has been repeatedly noted by commentators on mathematics, and I discuss it here; the other, oddly, is (pretty much) overlooked; I discuss it in 6.3.

It's been widely observed, in contrast to other kinds of conformity that *really do* have their source in social forces, that one finds in mathematical practice *nothing like* the variability found in cuisine, clothing, or metaphysical doctrine. There are examples of deviant computational practices: Babylonian fractions or the one-two-many form of counting; but overall empirical evidence for the possibility of deviation from standard mathematical practice—at least until the twentieth century—is meager.

Two points. First, as Kripke and others have noted (in the wake of the later Wittgenstein),[2] it's easy to *design* thought experiments in which people, impervious to correction, *systematically* follow rules differently from us. Despite the ease of *imagining* people like this, they're not *found* outside philosophical fiction. One *does* meet people who can't grasp rules at all—but that's different.

---

[2] E.g., Kripke 1982, Bloor 1983, and, of course, Wittgenstein 1953.

In rule-following thought experiments someone is portrayed who seems to follow a rule but who also understands "similar," so that she "goes on" differently from us. (After being shown a finite number of examples of sums, she sums new cases as we would until she reaches a particular border—pairs of numbers over a hundred, say—whereupon she sums differently, in some systematic way, while still claiming she's doing the same thing. This really is different from people who don't grasp generalization at all.)

Despite the absence of *empirically real* examples of alternative rule following (in counting or summing), such *thought experiments* are used to press the view that it's (purely) *social factors* that induce mathematical consensus. Given my remarks (in 6.1) about *appropriate* empirical methods for recognizing *real* options in group practices, such a claim—to be empirically respectable, anyway—can't batten on thought experiment alone; it needs an analysis of social factors that arise in *every* society that counts or adds—and that force humans to agree to the same numerical claims. The social factors pointed to, however—childhood learning, for example—are shared by almost every other group practice (diet, language, cosmetics, and so on) that show great deviation across groups. That is, even when systematic algorithmic rules (such as the ones of languages or games) govern a practice, that practice still drifts over time—unlike, as it seems, the algorithmic rules of mathematics.[3]

One possible explanation for this[4] is that practical exigencies exclude deviant rule-following mathematics: Someone who doesn't add as we do can be exploited—in business transactions, say. And so it's thought that deviant counting would die quickly. But this is sociologically naïve because even if the dangerousness of a practice did cause its quick demise, this wouldn't mean it couldn't emerge to begin with and leave evidence in our historical record of its temporary stay; all sorts of idiotic and quite dangerous practices (medical ones, cosmetic ones, superstitious ones) are *widespread*. Even *shallow* historical reading exposes a plethora of, to speak frankly, pretty dumb activities that (i) allowed exploitation of its practitioners (*and* helped shorten their lives), and yet (ii) didn't require too much insight on anybody's part to be *seen* as both pretty dumb and pretty dangerous. There's no shortage of such practices *today*—as the religious right and the raw-food movement, both in the United States, make clear. So it's hard to see why there can't have been really dumb counting practices that flourished (by, for example, exploiting the rich vein of number superstition that we *know* existed), and then died out (along with the poor fools practicing them).

Another way to deny the apparent sociological uniqueness of mathematical practice is the blunt response that mathematical practice *isn't* unique; there *are* deviant mathematical practices: Consider, instead of *counting*

---

[3] By "drift" I mean changes in the rules and practices that are not matters solely of mere augmentation, but of elimination in some cases. Mathematics is always being augmented; by specifically denying "drift" to it, I'm noting that such augmentation is overwhelmingly monotonically increasing.

[4] See, e.g., Hersh 1997, 203.

variants, the development of alternative mathematics—intuitionism, for example, or mathematics based on alternative logics (e.g., paraconsistent logics). Aren't *these* examples of mathematical *deviancy* every bit as breathtakingly different as all the things people willingly put in their mouths (and claim tastes good and is good for them)?

Well, *no*. What should strike you about "alternative mathematics"—unless blinded by an a priori style of foundationalism, where a specific style of mathematical proof (and logic), and a specific subject matter, are definitional of mathematics—is that such mathematics is *mathematics as usual*. One mark of the ordinariness of the stuff is that contemporary mathematicians shift in what they prove results about: They practice one branch of "classical" mathematics, and then try something more exotic—if the mood strikes. Proof, informal or formal, looks like the same thing (despite principles of proof being *severely* augmented or diminished in such approaches).

Schisms among mathematicians, prior to the late nineteenth century, are even *shallower* than this. That differences in methodology historically proved divisive can't be denied: Differences in the methodology of the calculus, in England and on the continent, for example, *retarded* mathematical developments in England for more than a century. Nevertheless, one finds British mathematicians (eventually) adopting Continental approaches to the calculus, and doing so because they (eventually) recognized that the results they wanted, and more generally, the development of the mathematics surrounding the calculus, were easier given Continental approaches. British mathematicians didn't deny the cogency of such results on the grounds that the methods that yielded them occurred in a "different (incomprehensible) tradition."

6.3 Eliminating Mistakes in Mathematics

Let's turn to the second (unnoticed) way that mathematics shockingly differs from other group practices. Mistakes are ubiquitous in mathematics. I'm not just speaking of the mistakes of professional—even brilliant—mathematicians, although notoriously they make many mistakes.[5] I'm speaking of ordinary people: They find mathematics hard—harder, in fact, than just about any other intellectual activity they attempt. What makes mathematics difficult is (i) that it's so easy to blunder in; and (ii) that it's so easy for others (or oneself) to see—when they're pointed out—that blunders have been made. (In other words, what makes mathematics hard is both how easy it is to make mistakes and how difficult it is to hide them.)

So what? This is where it gets cute. When the factors forcing behavioral consensus are genuinely social, mistakes lead to new practices—for two reasons at least: first, because the social factors imposing consensus can be blind to details about the behaviors enforced; they're better at imposing uniformity

---

[5] This is especially stressed in the "maverick tradition" to repeatedly hammer home the point that proof doesn't confer "certainty."

of behavior than at pinning down exactly which uniform behavior the population is to conform to. If enough people make a certain mistake, and if
enough of them pass the mistake on, the social factors enforcing consensus
continue doing so despite the shift in content. Social mechanisms imposing
conformity are good at synchronic enforcement; they're worse at diachronic
enforcement. (Thus what's sometimes described as a "generation gap.")[6]

The second reason is that the ability of social forces to impose conformity
often turns on the successful psychological internalization of social standards;
but if such standards are imperfectly internalized (and *any* standard—however
simple—can be imperfectly internalized), then the social standards themselves
can evolve, because, in certain cases, nothing else fixes them. Two examples are,
first, the drift in natural languages over time: This is often because of systematic
mishearings by speakers or interference phenomena (among internalized linguistic rules), so that certain locutions or sounds drop out (or arise). The
second example is when an external standard supplementing psychological
internalization of social standards is operative, and is *taken to* prevent drift—for
example, the holy books of a religious tradition. Notoriously, such things are
open to *hermeneutical drift*: The subject population reinterprets them (often
inadvertently) because of changes in language, "common sense," and therefore
changes in their (collective) view of what a given lawmaker (e.g., God) "obviously" had in mind.

In short, although every social practice is easy to blunder in, it's not at all
easy to get people to recognize or accept that they've made mistakes (and
therefore, if enough of them make a certain sort of mistake, it's nearly impossible
to restore the practice as opposed to—often inadvertently—starting a new one).

These remarks about mistakes don't imply that conscious attempts to
change traditions aren't effective: Of course they often are. But mathematical
practice resists willful (deliberate) change too. A dramatic case of a conscious
attempt to change mathematical practice that failed (in large part because of
incompetence at the standard fare) is Hobbes.[7] Another informative failure
is Brouwer, because Brouwer was anything but incompetent at the standard
practice.

Notice the point: Brouwer wasn't interested in developing *more* mathematics, nor were (and are) subsequent kinds of constructivists; he wanted to
change the practice. But he only succeeded in developing more mathematics,
not in changing that practice (as a whole). This makes Hilbert's response to
Brouwer's challenge misguided, by the way, because Hilbert's response was
also predicated on (the fear of) Brouwer's inducing a change in the practice.
This is common: Fads in mathematics often arise because someone (or a
group, e.g., Bourbaki) thinks that some approach can become *the* tradition of

---

[6] Consider school uniforms. All sorts of contingent accidents cause mutations in such uniforms; but that (at a time) the uniforms should be, um, *uniform* is a requirement. (It's very common for a population to slowly evolve its culinary practices, dress, accent, religious beliefs, etc.,
without realizing that it's doing so.)

[7] See Jesseph 1999.

mathematics—the result, invariably, is just (additional) mathematics. A related (sociological) phenomenon is the mathematical kook—there are enough of these to write books about.[8] Only a field in which the recognition of mistakes is extremely robust can (sociologically speaking) successfully marginalize so many otherwise competent people without standard social forms of coercion, e.g., the threat of prison.

So (to recapitulate) mistakes in mathematics are common, and yet mathematical culture doesn't splinter because of them, or for any other reason;[9] that is, permanent competing practices don't arise as they do with other socially constrained practices. This makes mathematics (sociologically speaking) *very odd*. Mathematical standards—here's another way to put the point—are robust. Mistakes *can* persevere; but mostly they're eliminated, even if *repeatedly* made. More important, mathematical practice is *so* robust that even if a mistake eludes detection for years, and even if many later results presuppose that mistake, this won't provide enough social inertia—once the error *is* unearthed—to prevent changing the practice back to what it was originally: In mathematics, even after lots of time, the subsequent mathematics built on the "falsehood" is repudiated.[10]

This aspect of mathematical practice has been (pretty much) unnoticed, or rather, misdescribed, and it's easy to see why. If one focuses on other epistemic issues, scepticism say, one can confuse the rigidity of group standards in mathematics with the possibility of certainty: One can claim that, if only sufficiently careful, really attending to each step in a proof, carefully analyzing proofs so that each step immediately follows from earlier ones, dutifully surveying the whole repeatedly until it can be intuited in a flash, then one can rig it so that—in mathematics, at least—one won't make any mistakes to begin with: One can be totally certain. (And then one can make this an epistemic

---

[8] For example, Dudley 1987.

[9] Philip Kitcher, during a discussion period on Nov. 21, 2002, after the presentation of a talk this chapter is based on, urged otherwise—not with respect to mistakes, but with respect to conscious disagreement on method: He invoked historical cases in which mathematicians found themselves arrayed oppositely with respect to methodology—and the suggestion is that this led to schisms which lasted as long (comparatively speaking) as those found among, say, various sorts of religious believers: One thinks (again) of the controversy over the calculus, or the disputes over Cantor's work in the late nineteenth century. But what's striking—when the dust settles, and historians look over the episodes—is how nicely a distinction may be drawn between a dispute in terms of proof procedures and one in terms of admissible concepts. The latter sort of dispute allows a (subsequent) consistent pooling of the results from the so-called disparate traditions; the former doesn't. Thus there is a sharp distinction between the (eventual) outcome of disputes over the calculus, and (some of) those over Cantor's work. The latter eventually flowered into a dispute over proof procedures which proved irresolvable in one sense (the results can't be pooled) but not in another. See 6.6. Also see 7.3 and 9.1, where I use this and other considerations to (eventually) argue that relatively fixed first-order proof procedures are what's behind ordinary mathematical practice—at least in what I (6.5) call "mature mathematics."

[10] Contrast this with our referential practices: Evans (1973, 11) mentions that (a corrupt form of) the term "Madagascar," applied to the African mainland, was mistakenly taken to apply to an island (the island we *currently* use the term to refer to). Our discovery of this error doesn't affect our current use of the term "Madagascar"—the social inertia of our current referential practice trumps social mechanisms for correcting dated mistakes in that practice.

requirement on all knowledge—and offer recipes on how to carry it off. Entire philosophical traditions start this way.)

But there are a number of, er, mistakes in this Cartesian line. First, it's a robust part of mathematical practice that mistakes are found and corrected. Even though the practice is therefore fixed enough to rule out deviant practices that would otherwise result from allowing such "mistakes" to change that practice, this won't imply that psychologically based certainty is within reach. For it's compatible with the robustness of our (collective) capacity to correct mathematical mistakes that some mistakes are still undetected—even old ones.

Apart from this, the psychological picture the Cartesian recipe for certainty presupposes is inaccurate. It's very hard to correct your own mistakes, as *you* know, having proofread your own work in the past. And yet (gallingly), someone else often sees your mistakes at a glance. This shows that the Cartesian project of gaining certainty all alone, a strategy crucial for Descartes' demon-driven epistemic program, is quixotic.[11]

Notice, however, that the Cartesian view would explain, if it were only true, how individuals can disagree on an answer, look over each other's work, and then come to agree on what the error is. (They become CERTAIN of THE TRUTH, and THE TRUTH is, after all, THE SAME.) Without this story, we need to know what practitioners have internalized (psychologically) to allow such an unnaturally agreeable social practice to arise.[12]

To summarize 6.2 and 6.3: What seems odd about mathematics as a social practice is the presence of substantial conformity on the one hand, and yet, on the other, the absence of (sometimes brutal) social tools to induce conformity that routinely appear among us *whenever* behavior really is socially constrained. Let's call this "the benign fixation of mathematical practice."

6.4 Attempts to Explain the Benign Fixation of Mathematics (That Fail)

The benign fixation of mathematical practice needs an explanation. And (it should be said) Platonism is an appealing one: Mathematical objects have their

---

[11] One of the ways Newton is so remarkable is that he did so much totally on his own, by obsessively going over his own work. See Westfall 1980.

[12] Unlike politics, for example, or any of the other numerous group activities we might consider, mathematical agreement isn't coerced. Individuals can see who's wrong; at least, if someone is stubborn, others (pretty much *all* the competent others) see it. Again, see Jesseph 1999 for the Hobbesian example. Also recall Leibniz's fond hope that this genial aspect of mathematical practice could be grafted onto other discourses, if we learned to "calculate together." By contrast, Protestantism, with all its numerous sects—in the United States, especially—is what results when coercion isn't possible (because deviants can, say, move to Rhode Island). And much of the history of the Byzantine Empire with its unpleasant treatment of "heretics" is the normal course of events when there's no Rhode Island to escape *to*. It's sociologically very surprising that conformity in mathematics isn't achieved as in these group practices. Imagine—here's a dark Wittgensteinian fable—we tortured numerical deviants to force them to add as we do. (Recall, for that matter, George Orwell's *1984*.)

properties necessarily, and we perceive these properties (somehow). Keeping our (inner) eye firmly on mathematical objects keeps mathematical practice robust (enables us to find mistakes). The problem—as the literature makes clear—is that we can't explain our epistemic access to the objects so posited.[13]

One might finesse a bit: demote Platonic objects to socially constructed items (draw analogies between numbers and laws, language, banks, or Sherlock Holmes). Address the worry that socially constructed nonmathematical objects like languages, Mickey Mouse, or laws, *evolve over time* (and that their properties look conventional or arbitrary), by invoking the *content* of mathematics (mathematical rules have content; linguistic rules are only a "semi-transparent transmission medium" without content). And, claim that such content makes mathematical rules "necessary."[14]

This won't work: We can't bless necessity upon anything we'd like by chanting "content." Fictional terms have content too—that doesn't stop the properties attributed to such "things" from evolving over time; socially constructed objects are *our* objects—if we take their properties to be fixed, that's something we've (collectively) imposed on them. It's a good question why we did this with mathematical terms and not with other sorts of terms.

If socially constructed objects are stiffened into "logical constructions" of some *fixed* logic plus set theory (say), this won't solve the problem either: One still must explain how logic (of whatever sort) and set theory accrue social rigidity (why won't we let our set theory and logic change?).

There is no single explanation for the benign fixation of mathematical practice because, as with any group practice, even if that practice retains its properties over time, that doesn't show that it has those properties (at different times) for the same reasons. Mathematical practice, despite its venerable association with unchanging objects, is an historical entity with a long pedigree; and so the reasons for why the correction of mistakes, for example, is robust in early mathematics aren't the same reasons for that robustness now.

6.5 *Mimicking Algorithms*

So now I'll discuss a number of factors, social and otherwise, and speculate how (and when) they contributed to the benign fixation of mathematical practice. The result, interestingly, is that benign fixation is historically contingent (and complex). That's a surprise for apriorists, but not for those of us who long ago thought of mathematics as a *human* activity.

---

[13] Current metaphysics robs Platonism of respectability. Judiciously sprinkle mysticism among your beliefs, and the perceptual analogy looks better; surreptitiously introduce deities to imprint *true* mathematical principles in our minds, and the approach also works. Explicitly deny all this, and Platonism looks *bizarre*.

[14] See Hersh 1997, 206. I deny that (certain) socially constructed objects, mathematical objects, and fictional objects exist *in any sense at all*. See Azzouni 2004a. Nominalism, though, won't absolve me of the need to explain the benign fixation of mathematical practice. On the contrary.

Let's start at the beginning: the historical emergence of mathematical practice (primarily counting and sums), and as that practice appears today among people with little or no other mathematics. Here it's appropriate to consider the role of "hard-wired" psychological dispositions. There seem two such relevant kinds of disposition. The first is a capacity to carry out algorithms, and—it's important to stress this—this is a species-wide capacity: We can carry out algorithms and teach each other to carry out (specific) algorithms *in the same way*. That's why we can play games with each other (as opposed to *past* each other), and why we can teach each other games that we can then play alone in the same way (e.g., solitaire).

I can't say what it is about us—neurophysiologically I mean—that enables us to carry out algorithms the same way; no one can (yet). It's clear that some of us are better at some algorithms than others (think of games and how our abilities to play them varies)—but what's striking is that those of us who are better aren't, by virtue of that, in any danger of being regarded as doing something else.

In describing us as able to "carry out" the same algorithms, I don't mean to say that we're *executing* the same algorithms (otherwise our abilities to carry out algorithms wouldn't differ in the so many ways that they do). I understand "executing an algorithm," as doing (roughly) what a Turing machine does when it operates. Perhaps humans do something like that with some algorithm(s) or other (but, surely, different humans execute different algorithms). In any case, when we learn arithmetic, for example, we're not learning to execute any of the numerous (but equivalent) algorithms that officially characterize arithmetic operations; instead what we're learning is what a particular algorithm is, and how to imitate its result—or at least, some of its results—by actions of our own. So when I describe us as "carrying out" an algorithm **A**, I mean that we're imitating it by doing something else **B**, not by executing *it*.

A way to see the point is to notice that our learning such algorithms enjoys an interesting flexibility: We not only (apparently) acquire and learn new algorithms, but we can get better at the algorithms we've already learned by practicing them. In addition, it should also be noted that, usually, mathematicians don't execute the algorithms they're officially deriving results from; they shortcut them.[15]

These considerations suggest that we don't—probably can't—execute the official algorithms we're carrying out; we're executing other algorithms instead that imitate the target algorithm (and over time, no doubt, different ones are used to do this); and this neatly explains why we can improve our abilities, by practice, to add sums, carry out other mathematical algorithms, and to win games (for that matter).

---

[15] Our ability to imitate algorithms flourishes into mathematical genius (in some individuals, anyway); for the mathematician, as I said, never (or almost never) figures out what an algorithm (proof procedure, say) will yield by executing that algorithm directly. Ordinary mathematical proof—its form, I mean—already shows this. See chapter 7.

Having said this, I must stress that I'm speculating about something that must ultimately be established empirically. So (of course) it could turn out that I'm just wrong, that we really do execute (some of) these algorithms (or, at least, that some subpersonal part of us does), and that we don't imitate them via other algorithms that we execute. The neurophysiologists, in the end, will tell us what's what (if it's possible for anyone to tell us this, I mean): I'm betting, however, on my story—as I said, it explains our algorithmic flexibility, and our capacity to make and correct mistakes, in a way that a story that requires us to actually execute such algorithms doesn't.[16]

I'm also unable to say—because this too is ultimately a matter of neurophysiology—how general our capacity to mimic algorithms is; that we can now (since Turing and others) formulate in full generality the notion of mechanized practice—algorithm—doesn't mean that we have the innate capacity to "carry out"—imitate—the results of any such algorithm whatsoever. Our capacity to imitate algorithms may be, contrary to (introspective) appearance, more restricted than we realize.

A (species-wide) capacity to imitate the execution of algorithms in the same way *doesn't* explain the benign fixation of mathematical practice. This is because that robustness turns on conserving the official rules governing mathematical objects, and a group ability to imitate algorithms the same way won't explain why a practice doesn't evolve by changing the applicable algorithms altogether—in just the way that languages, which involve algorithms too, evolve.

A second innate capacity that should be attributed to us is a disposition to execute certain specific algorithms.[17] I'm still thinking here primarily of our (primitive) ability to count and handle small sums. My suggestion is that primitive numerical practices are (pretty much) the same not because of sociological factors that constrain psychologically possible variants but because of fixed innate dispositions.

Don't read too much into this second set of dispositions since they're also too weak to explain the benign fixation of mathematical practice: They don't extend far enough. They're not rules that apply to, say, any counting number whatsoever. These dispositions are, I suspect, very specific: They may facilitate handling certain small sums by visualizing them or manipulating (mental) tokens in certain ways. Consequently there is no reason to think that such dispositions enable the execution of (certain) algorithms

---

[16] The last three paragraphs respond to a line of questioning raised by David Albert; my thanks for this. I should add that the neurophysiological evidence that's beginning to emerge supports my views. See notes 4 and 5 of 8.2 for indications of this.

[17] Caveat: Given my earlier remarks about the empirical nature of my speculations about how we imitate the execution of algorithms, I'm not sure I'm now describing a second capacity. Our ability to imitate algorithms—in general—needn't be a general ability to execute algorithms because, as just said, we don't "carry out" algorithms by directly executing them. What we do, perhaps, is apply a quite specific algorithm or set of algorithms to official algorithms that we want to carry out, a process that enables us to extract (some) information about any algorithm (once we've psychologically couched it a certain way).

so that we can count as high as we like, add arbitrarily large sums, and so on.[18]

6.6 How Mature Mathematics Fixes Mathematical Practice

Such innate dispositions as I've described, although they explain why the independent emergence of counting and summing among various populations always turns out the same, won't explain why, when mathematics becomes professionalized—in particular, when informal deduction is hit upon by the ancient Greeks—benign fixation continues, rather than mathematical practice splintering.

I introduce something of a sociological idealization, which I'll call "mature mathematics" and that I'll describe as arising somewhat before Euclid and continuing until the beginning of the twentieth century.[19] Several factors conspire to benignly fix mature mathematical practice.

The first is that pretty much until the twentieth century, mathematics came with intended empirical domains of application (from which mathematical concepts so applied largely arose). Arithmetic and geometry, in particular, have obviously intended domains of application. These fixed domains of application to some extent prevent drift in the rules governing terms of mathematics—in these subjects so applied, anyway—because successful application makes us loath to change successfully applied theorems if that costs us applicability.[20]

But something more must be going on with mathematics, as a comparison with empirical science indicates. For the history of empirical science (physics, in particular) shows that drift occurs without the intended application of the concepts and theories vanishing as a result. Newtonian motion, strictly speaking, occurs only when objects don't move. But its approximate correctness suffices for wide and successful application. Furthermore, the application of mathematics—geometry, especially—always involves (some) approximation because of the nature of what geometric concepts are applied to (in particular, fuzzily drawn figures).

---

[18] Thus I haven't (entirely) deflected Kripkean attacks on the dispositionalist approach to rule following. And: On my view dispositions have only a partial role in the benign fixation of mathematical practice. See the concluding remarks of 8.6.

[19] Twentieth-century mathematics isn't *mature*? Well, of course it is; but I'm arguing that it's different in important respects that require distinguishing it (sociologically, anyway) from what I've called "mature mathematics." Maybe—taking a nomenclatural tip from literature studies—we can call it "post-mature mathematics." On the other hand, maybe we'd better not. I'll contrast "mature" mathematics with "contemporary" mathematics.

"Sociological *idealization*" because aspects of mathematical practice that are present in "mature" mathematics (they're in Euclid) and that continue to play an important role in contemporary mathematics don't fit my official characterization of mature mathematics. I touch on this in 6.7.

[20] The ancient Greeks, it has been pointed out more than once, were disdainful of "applied mathematics." Yes, but that disdain is compatible with what I've just written. The view, for example, that the empirical realm is a copy of the mathematical realm both determines the intended empirical domain in the sense meant and simultaneously demeans the intellectual significance of that domain.

At work fixing mathematical practice beyond the drift allowed by successful (but approximate) application is a crucial factor: informal proof, or deduction. It's no doubt debatable exactly what's involved in informal proof, but in mature mathematics it can safely be described as this:[21] a canonization of logical principles, and an (open-ended) set of additional (mathematical) principles and concepts that (i) (partially) characterize subject matters with intended domains of application, (ii) are tractable at least insofar as we can, by their means, prove new unanticipated results, and (iii) grow monotonically over time.

The need for tractable mathematical proofs drives the existential commitments of mathematics, and in particular, drives such commitments away from objects characterized (empirically) in the intended domains of application. I've described this process in two case studies elsewhere, and won't dwell further on it now.[22] But the particular form that mathematical posits take now contributes several ways to benign fixation.

First, ordinary folk practices with empirical concepts allow those concepts to drift in what we can claim about them, and what they refer to, without our taking ourselves—as a result—to be referring to something new. If we discover that gold is actually blue, we describe that discovery in exactly those words (and not as a discovery that there is *no* gold).[23] By taking mathematical posits as empirically uninstantiated items, we detach mathematical language from this significant source of drift.

Second, once mathematical posits are taken as real but sensorily unavailable items that provide truths successfully applicable to empirical domains, philosophical concerns arise about both the nature of such truths and how such truths are established. For a number of reasons—mostly involving various prejudices about truth[24]—the conventionalist view that mathematical truths are stipulated and that mathematical objects exist in no sense at all isn't seen as tenable (or even *considered*), and a view of eternal and unchanging mathematical objects carries the day instead.[25]

---

[21] Recall, however, the second paragraph of note 19.

[22] See Azzouni 2004a and 2004b. The special qualities that a set of concepts, and principles governing them, needs for amenability to mathematical development—qualities that empirically derived notions and truths about them usually lack—may be found in Azzouni 2000a.

[23] I'm alluding here to the thought experiments of Putnam 1975b. See Azzouni 2000b, especially parts III and IV. There are subtleties and complications with this view of empirical terms, of course; but they don't affect the points being made now.

[24] Prejudices such as: (i) we can't have truths about things that don't exist—the truthmaker assumption—and (ii) even if we could have such truths, they wouldn't prove as empirically useful as mathematical truths have proven to be.

[25] Philosophical views about which positions are sensible and which aren't can't be ignored in any sociological analysis of why a group practice develops as it does. There are some, no doubt, who take philosophical views as mere ideology, as advertising for other more substantial social motives (e.g., professional or class interests). I can't see how to take such a position seriously, especially if it's the sociology of knowledge practices (of one sort or another) that are under study. What looked (or didn't look) philosophically respectable, I claim, had (and has) a profound impact on mathematical practice. It may be a mistake to search for that effect in the theorem-proving practices of the ordinary mathematician, but, in any case, as this chapter illustrates, I locate it in, as it might be

Of course, such an eternalist view of mathematical objects doesn't, all on its own, eliminate the possibility of a mathematical practice that allows drift in what we take to be true of mathematical objects: We could (in principle) still allow ourselves to be wrong about mathematical objects, and be willing to change the axioms governing them as a result. Imagine this thought experiment: a possible world much like ours, except that we discover nonEuclidean geometry centuries earlier, and because of the curvature of space in that world, its applicability is much more evident than in the actual world. In *that* world, we decide that Euclidean geometry is wrong; that is, we take ourselves to have been wrong about geometric abstracta—there are no abstracta that obey Euclidean axioms. This attitude is compatible with a view of mathematical abstracta as eternal, unchanging, and so on. What prevented such a view from emerging among *us*, I claim, is the relative *late* discovery of nonEuclidean geometry. I touch on this in 6.7; but my view is that had (one or another) nonEuclidean geometry emerged in ancient times, and had Euclidean geometry proved useless in its intended domain of application (in comparison to nonEuclidean geometry), Euclidean geometry would have been supplanted by nonEuclidean geometry; we would have taken ourselves to have been wrong about geometry and would have changed the basic axioms of what we called *geometry* to suit.[26]

I've stressed how intended domains of application helped to benignly fix mathematical practice; the (implicit) canonization of logical principles is just as essential. Had there been shifts in the (implicit) logic, then we would have found ourselves—when considering early mathematics—in exactly the same position as modern Greeks if they try to read ancient Greek on the basis of their knowledge of the contemporary stuff: incomprehension. In addition, shifts in the implicit logical principles utilized would have led to incompatible branchings in mathematical practice because of (irresolvable) disagreements about the implications of axioms and the validity of proofs.

6.7 How Contemporary Mathematics
Fixes Mathematical Practice

In order to motivate my discussion of how twentieth-century (and twenty-first century) mathematics differs from what came before, I need to amplify my claim about the (tacit) canonization of logical principles in mature

---

described, the general framework of how mathematics operates as a subject matter (in particular, in how it's allowed to change over time).

[26] This would have happened, in part, because of an implicit metaphysical role for mathematical objects in the explanation for why that mathematics applies to its intended empirical domain—recall the resemblance doctrine mentioned in note 20. But I attribute the fact that the late emergence of nonEuclidean geometry didn't supplant Euclidean geometry in part to the change, already in place, of "mature" mathematics into "contemporary" mathematics. I guess I'm hypothesizing a "paradigm shift" although I don't much like this kind of talk. It seems that Kline (1980) is sensitive to some of the changes from mature to contemporary mathematics, although he takes a rather darker view of the shift than I do.

mathematics. Contemporary discussions of Frege's logicist program, and the *Principia* program that followed it, often dwell—quite melodramatically—on paradox; and the maverick animus toward such projects focuses on the set-theoretic foundationalism that's taken to have undergirded both the ontic concerns and the obsession with rigor proponents of such programs expressed.[27] But this focus obscures what those projects really showed. Certainly they showed nothing about the (real) subject matter of mathematics (I rush to say), for that's proven to be elusive in any case. Ways of embedding systems of mathematical posits in other systems is so unconstrained, ontically speaking, that it has inspired structuralist views of that ontology.[28]

However, a very good case can be made that the logic of mature mathematics was something (more or less) equivalent to first-order predicate logic, and that this was a nontrivial thing to have shown.[29] Evidence for this claim is that the project of characterizing (classical) mathematics axiomatically in families of first-order classical systems succeeded.[30] What shows it isn't a trivial matter to so succeed is that, in fact, much of twentieth-century mathematics can't be so axiomatically characterized.

What's striking about this success is that the classical logic that's the algorithmic skeleton behind informal proof remained tacit until its (late) nineteenth-century *uncovery* (I coin this word deliberately). But, as the study of ever-changing linguistic rules shows, implicit rules have a slippery way of mutating; in particular, a general rule (at a time) can subsequently divide into a set of domain-specific rules, only some of which are retained.[31] The logical principles implicit in mathematical practice—until the twentieth century, however—remained the same topic-neutral ones (at least relative to mathematical subject matters). Such uniformity of logical practice suggests, as does the uniformity of counting and summation practices I discussed earlier, an at least partially "hard-wired" disposition to reason in a particular way.[32]

---

[27] In describing the complex history this way, I'm not necessarily agreeing with either the depiction or the attribution of these motives to later proponents of set-theoretical foundationalism.

[28] See, e.g., Resnik 1997.

[29] Why fix on first-order logic and not a higher-order (classical) logic, especially since it was a higher-order logic that historically arose first? See 7.3, 7.4, and 9.1 for a discussion of my reasons.

[30] Hersh (1997) and other mavericks deny this but offer only the (weak) argument that the project hasn't been carried out in detail for all the mathematics it was supposed to apply to. But why is that needed? The same grounds would show that Gödel's second incompleteness theorem hasn't been proven either; notice that it's irrelevant that the ordinary mathematician neither now, nor historically, couched any of his or her reasoning in such a formalism. This is because, as mentioned earlier, nobody carries out an algorithm by executing that algorithm—especially not gifted mathematicians who strategize proofs (and their descriptions) routinely, if they give proofs at all.

[31] See, e.g., S. R. Anderson 1988, esp. 334–35.

[32] A complication that (potentially) mars this otherwise appealing view of the implicit role of first-order logic in mathematics: The "logic" of ordinary language looks much richer than anything first-order predicate calculus can handle—notoriously, projects of canonizing the logic of anything other than mathematics using (even enrichments of) first-order predicate calculus have proved stunningly unsuccessful. This leaves us without analogous evidence that the tacit logic of natural language is (something similar to) first-order predicate calculus. But it would be very surprising if

This brings us to the points I want to make about twentieth-century mathematics. Contemporary mathematics breaks away from the earlier practice in two dramatic respects. First, it substitutes for classical logic (the tacit canon of logical principles operative in "mature" mathematics) proof procedures of any sort (of logic) whatsoever provided only that they admit of the (in principle) mechanical recognition of completely explicit proofs. That is, not only are alternative logics, and the mathematics based on such things, now part of contemporary mathematics; but various sorts of diagrammatic proof procedures are part of it as well; such proof procedures, which involve conventionalized moves in the construction of diagrams, needn't be ones easily replicated in language-based axiomatic systems of any sort.[33]

One factor that accelerated the generalization of mathematical practice beyond the tacit classical logic employed up until the twentieth century was the explicit formalization of that very logic. For once (a version of) the logic in use was made explicit, mathematicians could change it. Why? Because what's *conceptually central* to the notion of formal proof, and had been all along (as it had been operating in mature mathematics), isn't the presence of any particular logic or logical axioms of some sort, but only the unarticulated idea that something "follows from" something else. This is neat: Because (until the late nineteenth century) the logic was tacit, its particular principles couldn't have been seen as essential to mathematical proof *because they weren't seen at all*. What was seen clearly by mathematicians and fellow travelers (recall note 12) was the benign fixation of mathematical practice; but *that's* preserved by generalizing proofs to anything algorithmically recognizable, regardless of the logic used.[34]

---

the tacit logic of mathematics were different from that of ordinary language generally—especially given the apparent topic neutrality of that logic. There is a lot to say about this—indications of how I want to analyze the situation are found in 7.4, 9.1, and in the conclusion to part III.

[33] There's more to say about diagrammatic proofs, but not now. I've discharged the promissory note of the second paragraph of note 19, however: Diagrammatic proofs are in Euclid's elements (see Azzouni 2004b), and they continued to appear in mathematics during its entire mature phase—even though practices using them are only awkwardly canonized in a language-based theorem-proving picture of mathematical practice. The discussion of such items in contemporary mathematics is showing up in the literature on mathematical method. See, e.g., Brown 1999. I should stress again, however, that such practices require mechanical recognition of proofs; so they nicely fit within *my* (1994) characterization of mathematics as a structure of algorithmic systems. I should also add that diagrammatic practices within classical mathematics are clearly compatible with the tacit standard logic used there—they provide consistent extensions of the axiomatic systems they accompany (or so I conjecture)—and this is not true of the more exotic items (e.g., logics) invented in the twentieth century.

[34] Haim Gaifman (Nov. 21, 2002) has raised a challenge to the idea that contemporary mathematics can (genuinely) substitute algorithmic recognizability for the implicit logic of mature mathematics. For given the fact—aired previously—that mathematicians don't execute the actual algorithmic systems they prove results from, it must be that they rely on methods (modeling, adopting a metalanguage vantage point, etc.) which incorporate, or are likely to, the classical logic mathematicians naturally (implicitly) rely on. This suggests that if a mathematician were to attempt to *really* desert the classical context (and not merely avoid a principle or two—as intuitionists do), he or she would have to execute such algorithms mechanically; any other option would endanger the validity of the results (because the shortcuts used could presuppose inadmissible logical principles).

The second way that contemporary mathematics bursts out from the previous practice is that it allows pure mathematics such a substantial life of its own that areas of mathematics can be explored and practiced without even a hope (as far as we can tell) of empirical application.[35] This, coupled with the generalization of mathematical proof to mechanical recognition procedures (of any sort) allows a different way to benignly fix mathematical practice. For now branches of mathematics can be individuated by families of algorithmic systems: By (tacit) stipulation, one doesn't change mathematical practice; new mathematics is created by the introduction of new algorithmic systems (i) with rules different from all the others, and (ii) that aren't augmentations of systems already in use. Should such an invention be empirically applicable, and should it supplant some other (family of) system(s) previously applied to that domain, this doesn't cause a change in mathematics: The old family of systems is still mathematics and is still something profitably practiced (from the pure mathematical point of view). All that changes is the mathematics applied (and perhaps, the mathematics *funded*).

Notice that these reasons for the benign fixation of mathematical practice differ from those at work in mature mathematics. In particular, recall (from 6.6) my thought experiment about the much earlier discovery of nonEuclidean geometry in a nearby possible world; its discovery in our world, given when it happened, spurred on the detachment of mathematics (as a practice) from intended domains of application; but that was hardly something that it *started*. Mathematical development had already started to explode (in complex analysis, especially). But although intended domains of application were still exerting strong pressure on the direction of mathematical research, the introduction of mathematical concepts was no longer solely a matter of abstracting and idealizing empirical notions, as the notion of the square root of -1 makes clear all on its own. I claim (but this is something only historians of mathematics can evaluate the truth of) that this, coupled with a more sophisticated view of how mathematical posits could prove empirically valuable (not just by a "resemblance" to what they're applied to), and both of these coupled with the emergence of a confident mathematical profession not directly concerned with the application of said mathematics, allowed the birth of mathematical liberalism: the side-by-side noncompetitive existence of

---

It may be right that an adoption of a seriously deviant nonclassical logic for (some) mathematics requires *compete formalization*. I'm simply not sure: As the discussion in chapter 4 indicates, how much a regimentation can desert the vernacular is a sticky empirical matter. We *could*, of course, completely formalize our mathematics (based on a seriously deviant logic) and rely on computers for the results. This raises a tangle of issues that I can't analyze now. See Azzouni 2005.

[35] What about classical number theory? Well, I'm *not* claiming that "mature" mathematics didn't have subject matters, the exploration of (some of) which wasn't expected to yield empirical application; but numbers aren't the best counterexample to my claim since they were clearly perceived to have (intended) empirical applications. The contemporary invention and exploration of whole domains of abstracta without (any) empirical application whatsoever is a different matter. Consider, e.g., most of the explorations of set-theoretic exotica or (all of) degree theory. (None of this is to say, of course, that empirical applications can't arise later.)

(logically incompatible) mathematical systems. And what a nice outcome that was!

6.8 Concluding Remarks

Even if we (rightly) reject the idea that mathematical truth is a priori, that it's something we grasp by non-empirical methods, and therefore is something we have (in principle, anyway) access to total certainty about, and even if we substitute for this the idea that mathematical truths are not epistemically different in principle from empirical ones—that mistakes are possible, and that we often convince each other we're right about something by going at it as a group—there is still something special about how we agree about mathematical proofs. I've tried to show how this something special can't simply be a matter of the grasping of algorithms, because arguably the same is true of our ability to speak to each other—and yet the rules for languages drift in a way that the principles underlying proof in mathematics don't.

I've given a historical explanation for the benign fixation of mathematics—an explanation that shifts over time just as mathematics itself does. In this way, although the phenomenon of benign fixation looks the same, it turns out that the reasons for it have changed. Still, the centrality of algorithms remains a constant. The interesting question, therefore, is how exactly the presence of such algorithms manifests itself in ordinary mathematical proof—a practice that doesn't look syntactic in the slightest, but one in which the subject matter seems to play an ineliminable role in how the mathematician recognizes successful proof. I turn to an analysis of this question in chapter 7.

7

The Derivation-Indicator
View of Mathematical
Practice

7.1 Morton's Challenge

I once optimistically offered this thought experiment:

(Imagine that mathematical objects ceased to exist sometime in 1968. Mathematical work went on as usual. Why wouldn't it?)[1]

Adam Morton has suggested—contrary to my sanguine and parenthetical prophecy—that the result could (perhaps) be nothing short of apocalyptic. For mathematicians might lose their imaginative powers: suddenly be unable to speculate about mathematical objects—consequently, be unable to hypothesize possible results they could then try to prove. Instead, those mathematicians stubbornly intent on continuing their art in the wake of the brutal abstractacide just contemplated would have to resort to blindly constructing numerous detailed but boring proofs of utterly trivial results in the hopes that doing so day after day, year after year, might finally result in their groping their way to something interesting. Their new proofs would, it's true, have the virtue of *really* being mechanically checkable for validity; and this would raise the possibility that even though mathematicians were now blindly proving results about objects they could no longer intuit, they would still *recognize* an interesting result should one turn up. But alas, even this would now be impossible: For such a failure of imagination would prevent them from distinguishing interesting from uninteresting results. What makes a result interesting is its

[1] Azzouni 1994, 56.

illumination of the objects it's about, and mathematicians would now be unable to tell if such illumination had taken place. As a result, the great tradition of mathematics would be permanently sealed off from them. Attempts to read previous mathematics—even their own—would be similar to experiences of certain aphasics after a stroke; they wouldn't (now) understand what such proofs even *meant*. The terribly sad result: Mathematics would no longer be something *humans* could engage in.[2]

7.2 The Distinction between Proofs and Derivations

It's the purported threat of mathematical aphasia that I'm most concerned with in this chapter. To this end, I focus on a distinction found in Rav 1999 (11),[3] between a *proof*, which appears in "customary mathematical discourse" and has "an irreducible semantic content," and a *derivation*, which "is a syntactic object of some formal system." Derivations are seen by Rav as meaningless; such meanings as they are associated with (12) "are now shifted to the metalanguage." When speaking of proofs, he writes of "semantic elements, contextual relations and technical meanings." His aim, in making this distinction, is to challenge "the formalist"—someone taken to believe that (in principle, anyway) derivations can replace proofs altogether in mathematical practice. My sympathies to "formalism" don't extend this far, as this chapter will make clear.[4] My more immediate aim, however, is to connect this distinction to Morton's challenge by suggesting that the challenge implies that, without access to mathematical objects, the mathematician is restricted to derivations—access to *proof* requires access (on the mathematician's part) to the mathematical objects themselves (that one is proving things *about*). To cement a connection between Morton's challenge and Rav's distinction, however, one must take the "irreducible semantic contents" involved in *proof* to involve reference to *mathematical objects* of some sort.[5]

Rav doesn't indicate what relata, if any, he's willing to attribute to mathematical terms. It's clear, though, that he contrasts the semantically laden proof with the meaningless derivation to the detriment of the latter—at least as far as mathematical practice is concerned. My aim therefore is twofold. It's to show, first, that the purportedly formalist-resistant aspects of mathematical practice

---

[2] Personal communication on October 8, 2002. I've described Morton's variant on my thought experiment in my own way. I'm under the impression, incidentally, that he intended only to offer me an interesting thought experiment counter to mine, and wasn't expressing a commitment to any particular view about what his thought experiment implied, or even, in fact, a commitment to the claim that mathematicians *would* suffer a failure of imagination if mathematical objects "disappeared."

[3] Rav's distinction, or items close to it, is fairly widespread in the literature. See, e.g., DeMillo et al. 1979. It often occurs under the nomenclature: "informal"/"formal" proof.

[4] And, in fact, as I've already made clear in Azzouni 1994, e.g., 147–49.

[5] "Mathematical objects," though, is understood broadly: Perhaps such objects are *relations*, perhaps concepts. All I'm requiring of the view is that mathematical terms refer to *something* the correct grasping of which plays an ineliminable role in our understanding of mathematical proof.

that Rav focuses on can be captured by views of that practice which take mathematical objects as nonexistent. Whether, as a result, mathematical terms are "meaningless" is a question I evade because of a certain unclarity in what it's supposed to mean. In one sense of "meaningless," mathematical terms aren't meaningless for they are linked both to applications outside mathematics and to other terms in mathematics, specifically, to those in other mathematical systems.[6] Nor do I claim that mathematicians don't or shouldn't engage in mathematics via complicated "object-centered" psychological processes. Many mathematicians, that is, imagine themselves to be exploring properties of objects (of a certain kind) by means of mental representations (of them) that are manipulated (visually, kinesthetically, etc.) just as, say, a pitcher, in imagining how he or she might pitch a ball, imagines moving in certain ways, the ball moving a certain way as a result, and so on. All of this can be conceded, all of this can be seen as (perhaps) practically indispensable to successful mathematics, without any of it committing us to mathematical objects as actually existing in any sense at all.[7]

The second thing I want to show is that although it's true (on my view) that *proofs* in Rav's sense are indispensable to mathematical practice, this is compatible with the claim (also true in my view) that it's *derivations* in one or another *algorithmic system* that *underlie* a characteristic of mathematical practice: the *social conformity* of mathematicians about whether a proof is or isn't (should be, or shouldn't be) convincing. There is a strong tendency (in some circles) to argue that proof, in Rav's sense, is *only* what mathematicians find convincing—that mathematical proving is *merely* a "socially constructed" practice, where that phrase implies, that is, that there is *nothing more to say* (apart from how social factors induce uniformity in the behavior of a group of primates) about how it is that mathematicians so universally find themselves agreeing that a certain proof suffices to establish the result it purports to establish.[8] In describing a social practice as "merely" a socially constructed

---

[6] See 7.9 for more details. Also see Azzouni 1994, esp. 147–49. To direct-reference theorists, however, mathematical terms *are* meaningless (on views like mine), since they're taken to *refer* to nothing *at all*.

[7] The need for mental representations of things doesn't by itself imply that those things exist. A novelist can engage in quite subtle ratiocinations about characters; indeed, "object-centered" thinking of this sort may be indispensable to being a successful novelist; but that doesn't require ontic commitment to things imagined and represented. From the logical point of view "object-centered" thinking translates into objectual quantification—and *that* logical tool is available without any accompanying ontic commitments whatsoever. See Azzouni 2004a, 53–55.

[8] See DeMillo et al. 1979. See MacKenzie 2001 for a discussion of the sociological impact of the DeMillo et al. paper and for a discussion of this view of proof on the part of others. MacKenzie (251) quotes John Kershaw: "Proof is a typical Humpty Dumpty word, which means precisely what you want it to mean. . . . In practice proof seems to mean 'an argument that most practitioners in the field accept as valid,' no more and no less." I distinguish this claim about "proof"— which is wrong—from the following sentiment, which could be confused with it (and which is largely, although not entirely, right). DeMillo et al. (271) write: "It is a social process that determines whether mathematicians feel confident about a theorem." This isn't entirely right—as a matter of sheer sociological fact—because it's not true of *all* mathematicians with respect to *all* theorems; but in any case, the fact that it's right for, say, *most* mathematicians with respect to *most*

practice, one goes beyond the truism that mathematical practice is social[9] to the claim that *only* social *factors* are causally pertinent to the conformity exhibited by mathematicians.

In contrast, I take a proof to *indicate* an "underlying" derivation. *How* proofs do this is complicated, and I'll say more about it shortly. But the point to make *right now* is this: Because (i) derivations are (in principle) mechanically checkable, and because (ii) the algorithmic systems that codify which rules may be applied to produce derivations in a given system are (implicitly or, often nowadays, explicitly) recognized by mathematicians, it follows that if what makes proofs convincing are factors that lead to those proofs being correlative to such derivations, this suffices to explain why mathematicians are so good at agreeing with each other about whether some proof convincingly establishes a theorem.[10] Any approach focusing on (purely) social constraints on behavior can't explain how mathematicians agree on the status of proofs, as we saw in 6.1; and this is because social factors, even when they induce conformity at a time, can't (and don't) prevent *substantial* diachronic change.

7.3 The Derivation-Indicator View of Mathematical Practice; and Ruminations on Whether Such Derivations Must Be Construed as First-Order (at Least in Mature Mathematics)

My approach needs a name, and I want a straightforward one. Since, on this view, proofs are *derivation indicators*, I'll call it *the derivation-indicator view of ordinary mathematical proof* (the DI view). Even granting the objection just made to any approach that leans only on (purely) social factors to explain uniformity of behavior among mathematicians, one may worry about the role in mathematical practice attributed to derivations on the DI view. This view, as I've said, *doesn't* take mathematics as something in which the practice of providing proofs can be replaced by a practice in which only derivations are utilized; that's completely implausible. But the DI view may seem as bad: It, after all, claims that a proof of a theorem indicates a derivation of that theorem (in some informally specified algorithmic system or other). Why should this "derivation-revelation" view of mathematical practice be taken seriously?

One extraordinary piece of good evidence for it is the *success* of the early twentieth-century logicist program, as I mentioned in 6.7. Algorithmic

---

theorems *doesn't mean* it implies Kershaw's sentiment; nor does it falsify any of the claims made in this chapter or the next about "derivations."

[9] Truism? Sure—since there's hardly any human activity that isn't "social."

[10] This suffices only if the notorious rule-following problem doesn't infelicitate an acceptable story for how mathematicians can recognize and follow rules that doesn't itself fall back upon the very social factors (for behavioral conformity) that I deny can explain what needs explaining. However, I leave explicit discussion of the rule-following problem in all its details for another time. (I say *something* about it at the end of 8.6.)

systems, however, are restricted neither to a particular logic, a particular subject matter, nor even to an explicit language (as opposed to something diagrammatic or pictorial). What *is* required—and all that's required—is that derivations, however these be understood, are (in principle) mechanically recognizable. As suggested in 6.7, what I there called contemporary mathematics involves research into all sorts of systems; and thus (in principle, anyway), allows the indicating of derivations from all sorts of families of algorithmic systems.

I've suggested, however, that the families of algorithmic systems of the derivations so indicated that are still in the spirit of what I called (in 6.6) mature mathematics—mathematics that can be replicated with the resources of ZFC, say—are first-order ones.[11] So the suggestion isn't merely that families of first-order systems are good regimentations for classical mathematical proofs, but more strongly that in some sense such families of first-order systems are the items that were *really* indicated all along. But is such a specific assumption about the tacit logic of mature mathematics necessary? And, anyway, isn't it implausible since the logic that first emerged (at the hands of Frege), and put to use in the logicist project by Russell and Whitehead wasn't first-order? The latter item emerged later—first distinguished by Löwenheim (in 1915) and shortly thereafter championed by Skolem.[12]

If we restrict our attention to the historical data discussed in chapter 6—the benign fixation of mathematical practice, and the accompanying absence of drift in the notion of a successful proof—then strictly speaking, standard first-order logic *isn't* required as the tacit logic of mature mathematics. Let me state baldly what these aspects of the historical record require, and then (in this section, and in 7.4) I'll note some refinements: (i) the principles of proof—the logic—are (more or less) fixed for the duration of mature mathematics, (ii) those principles of proof are effective, and (iii) the concepts of mature mathematics, however, fluctuate in two senses. First, additional concepts steadily accrue to the subject matter—concepts governed by tacit and explicit principles which interact synoptically with what came before. Second, concepts, even when treated as identical to earlier ones, are steadily supplemented with additional principles that—strictly speaking—are independent of earlier principles, and thus make the whole package of principles—tacitly and explicitly governing a concept—ever more powerful.

The historical record motivates this much for the reasons we saw in 6.2 and 6.3: Unlike natural languages, where the syntax drifts, mathematical proofs are quite robust. Later mathematicians may find that tacit assumptions not previously expressed need explicit notice, but one doesn't find earlier proofs to be products of alien "reason," that we've subsequently evolved beyond. One factor that could enable the logic of mathematical reasoning—whatever it

---

[11] Many, if not most, contemporary mathematicians continue to work within a more powerful extension of mathematics as it emerged at the turn of the last century—all of which can be captured within first-order ZFC.

[12] See Moore 1980 for this history.

is—to remain implicit was *precisely* the fact that drift in the proof procedures didn't occur. What's striking about contemporary mathematics, carried out on the basis of alternative logics, is that they must be—more or less—formalized in order to see what is and isn't provable in them. Drift in the logic of (some research areas of) contemporary mathematics is reflected necessarily in the explicitness of the (alternative) logical principles used.

The official centrality of the mathematical object in mature mathematics— that the mathematician proves things about objects of a certain sort—allowed mathematical concepts to shift in content without recognition of the fact. As long as numbers are, explicitly or implicitly, seen as a kind of object, one can treat the consistent supplementation of the principles governing the concept of number as further discoveries about *numbers*. The natural thought then is that one is (all along) studying the same objects, and therefore, referring to the same objects, even if the principles governing these things shift.

This tension between the shifting of our mathematical concepts (by adding to their principles) and the perception that the subject matter hasn't changed is visible in current debates about whether mathematical proof is best captured by first- or higher-order logics. Some sacrifice the effectiveness of our notion of proof for categoricity: referential access to the mathematical objects at the cost of steady mutations in proof procedures. But such referential access is an illusion of the notation of second-order logic thus understood: a matter of fiat.[13] Thus my counter-suggestion that the historical record requires an effective (and relatively fixed) set of logical principles, and consequently, an evolution of mathematical concepts instead.[14]

7.4 A Programmatic Claim about the Relationship between the Logic We Employ in Mature Mathematics (and, Arguably, in Ordinary Life) and the Syntax of Natural Languages

If one looks (closely) at the grammar of ordinary languages, the evidence is good that the expressive powers of such languages outstrip first-order idioms: the presence of generalized quantifiers ("most," "approximately two hundred," "all but finitely many,"),[15] as well as apparent examples of branching[16] and second-order quantifiers.[17]

This, at least prima facie, suggests that the "logic" of natural language isn't first-order, or at least, that "the" logic in question is opportunistic and variable. Leaving aside tense, however, there is no reason why richer expressive properties—if they led beyond first-order logic in their theorem-proving

---

[13] See the discussion of this in Azzouni 1994, § 3.

[14] See 7.11, where I discuss this picture further.

[15] See the classic Barwise and Cooper 1981, and for a more recent survey, Keenan and Westerståhl 1997.

[16] See, e.g., Keenan and Westerståhl 1997, 880–82, and the citations therein.

[17] See, e.g., Boolos 1998b.

effects—wouldn't have emerged in mature mathematics in such a way as to have prevented capture (in first-order terms) of that mathematics.[18] The right picture of the logic of natural language is, I suspect, that despite the semantics the rich expressive resources of natural language require, at no syntactic level is the logic we use syntactically visible.[19]

The picture, therefore, is this: Specific logical relations between sentences—even if these, phenomenologically, have a semantic feel—are recognized by couching them in something effective. And this is regardless of the fact that the expressive capacity of our language, even as it occurs in mathematics, outstrips anything effective. Several points should be made about this claim. First, the invisibility of the actual principles we use—it's worth noting—is hardly atypical of mathematics. The history of mathematics, especially (early) results about numbers, occurs in ordinary language with virtually no addition of technical vocabulary; it takes place in a language that fully masks the algorithms enabling recognition of the results.

This brings us, however, to the second important point. This is that the algorithmic methods and terminology invented by the mathematician to facilitate computation and theorem proving, even when used to compute and show results available before the invention of said devices, needn't correspond to the (psychological) tools mathematicians used previously. Exactly what properties representations—of numbers, say—have in mathematicians' minds when they engaged in computations is an empirical question. It should not be assumed that terminological inventions, of Arabic notation, say, are making official and conscious the psychological representations of numbers that mathematicians had before that date.

Third, even if the logic of mature mathematics that eventually emerges is the first-order predicate calculus, it doesn't follow that the algorithmic methods psychologically available to mathematicians are themselves best represented as axiom systems in first-order languages. Rather, the algorithmic systems so available are likely to be ones that involve inference patterns that contain additional mathematical content, and within which what we currently take to be *logical* inference rules are implicit. Furthermore, they are likely to be topic-specific, in contrast to unadorned logical principles. In Azzouni 2005 I described these as inference packages and say more about them there.

A last point should be made about this. Although some numerical patterns of reasoning were made syntactically explicit by, say, Arabic notation—and "syntactically explicit" only requires, as I urged above, the mimicking of the numeral pattern of reasoning, not its actual exposure—which was invented

---

[18] One argument for something being lost in the desertion of higher-order logic—categoricity—I've already (long) attacked: See note 13 (but also see 9.1). Another argument turns on the inordinate length of (some) first-order derivations. I take this argument up in 7.6.

[19] That is, there is no syntactic level at which the grammar wears its logical properties on its sleeve, as happens in first-order logic where there are a set of syntactic manipulations that can be applied to sentences, and which are coextensive with what we perceive to be the implication relations among those sentences (i.e., the existence of completeness and consistency proofs). I discuss the implications of this claim in more detail in the conclusion to part III.

terminology that remained within the vernacular, effective proof procedures can be made explicit only by deserting ordinary language altogether—regimenting them in a formal language—and the possibility of doing *that* had to wait until the invention of such languages.[20] And this gives yet another reason for why the effective proof procedures of mature mathematics remained tacit for so long.

A qualification: There are formalisms that aren't syntactically first-order, but that have effective proof procedures. Nothing said so far has ruled these out as suitable reconstructions of ordinary mathematical reasoning. For my current purposes, as long as the effectiveness and relative fixity of the proof methods of mature mathematics—and the accompanying shifting of its conceptual content (without explicit acknowledgment)—are respected by the families of algorithmic systems containing the indicated derivations, any such families of algorithms will do. It's compatible with the practice of mature mathematics that (i) derivations were being indicated, but (ii) without practitioners being aware that their ability to agree on proofs was because of this, and (iii) that as a result, there is no fact of the matter—in certain respects—about *which* particular sort of derivations were being so indicated. See 7.7. So with these caveats in place, I'll continue to speak of the implicit logic of mature mathematics as being first-order.[21]

7.5 Derivations and Proofs: Some Preliminary Observations

I now want to illustrate how the apparent role of the mathematical term to refer to objects actually codes the various ways that a traditional proof can be correlated to one or another *derivation* within (one or another) algorithmic system. But before supporting this claim, I should make clear certain important aspects of *derivations* and *proofs*, as I use these terms. What should be stressed first is that, just as algorithmic systems aren't restricted to particular subject matters (any set of "concepts," nearly enough, can be mathematized), so subject matters (as understood in day-to-day mathematics) are not restricted to particular algorithmic systems. The standard interpretation of Gödel's theorem takes mathematical subject matters (such as the counting numbers) to go beyond any single algorithmic system (PA, for example); but ordinary mathematical practice, with its routine introduction of new concepts via new notation to facilitate proofs and understanding, already involves the routine

---

[20] Thus it's no historical accident, on my view, that Frege both (largely) discovered the logic tacitly underlying mature mathematics *and* championed an artificial language to exhibit that logic within.

[21] I revisit the question of the logic of mature mathematics in 9.1. I try to show there that additional considerations not raised here support the stronger thesis I've (temporarily) backed away from: that the logic of mature mathematics *really is* first-order. I also raise very different considerations for why the logic underlying mature mathematics should be taken to be the first-order predicate calculus in Azzouni 2005.

augmentation of systems. The only *logical* requirement on such augmentations (the only requirement apart from considerations such as that they allow us to prove theorems more easily—this can mean, of course, that they admit (practically speaking) a proof to be produced *at all*—or, allow us to provide a proof that's more illuminating, etc.), is that the result conserve previous systems. That is, results of earlier nonaugmented systems aren't disallowed when augmenting such systems with additional proof-enabling tools. As we'll see, this aspect of mathematical practice—the tying of algorithmic systems together—is absolutely *central* to mathematical practice, and it explains a lot that would otherwise need reference to mathematical objects (numbers, for example) to try to explain.

The other preliminary points are about the other half of Rav's distinction: *proof.* Here it's natural to invoke the notion of a "proof sketch."[22] (In line with my nomenclature, I use "derivation sketch.") A claim often made is that proofs are derivation sketches. We must be careful, though, because terms like "proof sketch" or "derivation sketch" are misleading. Those trained in logic are familiar with the distinction between the axioms and inference rules of a formal system, and the *derived* or *subsidiary* rules added to such a system. Such subsidiary or derived rules operate in one of two ways. The first is as additional rules for the construction of derivations; the second is as "metamathematical" rules used in proofs *about* derivations—in particular, proofs that certain derivations *exist.*[23] In either case, the rules are shown to be *conservative* in their proof-theoretic import: The proofs that such subsidiary rules are conservative are constructive insofar as they implicitly show how to produce a derivation of a particular result from a given derivation of a different result (in the second case) or how to produce a derivation that uses only the original axioms and inference rules of the system from a given derivation containing (in addition) subsidiary derivation rules (in the first case). In this sense, such rules are *local*: They implicitly provide operations on proofs (or derivations) that transform the former into derivations of the desired sort.

It may be that "derivation sketch" or "proof sketch" as commonly used in the literature is the informal analogue of the logical practice just described with subsidiary inference rules. Notice, in particular, that the metamathematical analysis so described (e.g., by Kleene) despite its being "metamathematics" doesn't involve *semantics*—the domain of metamathematical analysis is syntactic: Constructive relations among derivations are established, where such relations depend (only) on the derivation rules. The phrase "derivation sketch," fits this proof practice reasonably well (despite some slight terminological awkwardness with respect to the case where one transforms a derivation of one result into a derivation of another): The sketch can be "filled out" into a genuine derivation.

[22] See, e.g., Fetzer 1988. I also invoke them in Azzouni 1994, 158–59: "Proof sketches are best understood as extrasystemic, enthymemic arguments that proofs exist."

[23] For a rigorous illustration of a formal system, and the introduction of derived rules in the metamathematical approach, e.g., the deduction theorem, see Kleene 1971, especially chapter 5.

The proof practices of mathematicians, however, go beyond these derivation-sketch resources. When mathematicians give what I'll describe as "arguments that indicate derivations" (ADs),[24] the latter often seem to turn *essentially* on the subject matter that the derivations concern. As a result, a direct characterization of them in purely derivation-theoretic terms eludes us. Here is Barwise's (1989, 849) allusion to the phenomenon:

> There are many perfectly good proofs that are not modeled in any direct way by a formal proof in any current deductive system. For example, consider proofs where one establishes one of several cases and then observes that the others follow by symmetry considerations. This is a perfectly valid (and ubiquitous) form of mathematical reasoning, but I know of no system of formal deduction that admits of such a general rule. They can't, because it is not, in general, something one can determine from local, syntactic features of a proof.

Talk of symmetry considerations among mathematicians is loose, however, and there is, no doubt, more than one kind of thing going on when the phrase is invoked. Here are a couple of simple examples: Consider a set A with an equivalence relation defined on it. Given an element $a \, \varepsilon \, A$, we define an equivalence class as all those elements in A equivalent to a: $\mathrm{cl}(a) = \{x \, \varepsilon \, A \mid a \sim x\}$. It can be shown that any two equivalence classes are either disjoint or equal. Assume (to begin with) that they aren't disjoint: that there is an element contained in their intersection. One then assumes that element to be in one of the sets (choose one at random), and concludes that it must be in the other one. (This shows a one-way inclusion.) One then notes that the "argument is clearly symmetric,"[25] and so one has shown the identity of the two sets.

Where, exactly, is the symmetry here? Well, although one knows that equivalence relations are symmetric, the definition of an equivalence class places the element used to define the equivalence class on the right side of "$\sim$." And the transitivity condition is of this form: "$x \sim y$ and $y \sim z$ only if $x \sim z$," not, for example, "$x \sim y$ and $z \sim y$ only if $x \sim z$." Although the reversals one must engage in (at one or another place in the proof) are trivial, and justified by the equivalence relation on A being an *equivalence relation*, they must still be explicitly executed; and this makes the two proofs different in a way that's not amenable (directly) to a logical rule.

---

[24] Although I used the term "proof sketch" in my 1994, I *didn't* intend the nomenclatural implication that all that's needed to exhibit the derivation (in an algorithmic system) that the proof sketch indicates is for "missing steps" to be filled in. That this can't be right is already clear from standard proofs of the deduction theorem, since those operate on a (given) *entire* derivation to transform it into the desired derivation. But even this is too restrictive: The proof a mathematician gives may indicate a derivation in a way that doesn't turn on the applicability of transformations to *that proof* at all. One reason for this, as we'll see, can be because the proof relies implicitly on certain semantic characterizations of what the proof is about; these semantic characterizations, in turn, can be code for a rather different proof (referring, in part, to different *objects*) that the adept is simply assumed to be able to handle. Thus my adoption of the less attractive sounding terminology: "arguments that indicate derivations."

[25] Herstein 1975, 8.

Other elementary examples: Consider the real line, and recall the notions of lower bound, upper bound, greatest lower bound (infimum), least upper bound (supremum), of a subset of the real line. Also recall (theorem) that each nonempty set S bounded above has a supremum. The proof of its companion theorem (each nonempty set S bounded below has an infimum) is "symmetrical" to the supremum proof: One must reverse the inequalities in the first proof.[26]

The lesson of these examples is this: Proofs seen as "symmetrical" are often so seen because the *relations* and *predicates* involved in the proof are "symmetrical"; by judicious *replacement* of these relations and predicates with *other ones* (symmetrically related to the originals), one can syntactically imitate a given proof, and in this way provide a proof of its symmetrically related companion. These ADs go beyond "derivation sketches" as described above; but only in the following way: In cases like the deduction theorem one modifies a derivation step by step so that the result is a new derivation. Here too (at least in the examples given, and others like them), one also goes through the derivation and modifies it step by step so that the result is a new derivation. What prevents the approach from being *systematically* amenable to the metamathematical approach of, say, Kleene 1971, is that the tools enabling the transformations aren't logical axioms, but are topic-specific truths about specific relations and predicates (of the subject matter). Thus, the "fleshing out" image that the phrase "derivation sketch" conjures up still applies as well as it ever did; and it doesn't seem to immediately follow that intrinsically semantic considerations are involved—at least not in the sense that *mathematical objects* have, via such proofs, now made their appearance in mathematical practice. *Topic-specific facts* are relevant, but the way these are linked to predicates and relations (rather than objects) is what prevents systematic (explicit) logical expression of these facts— namely, as subsidiary proof procedures.

7.6 Why Second-Order Logic Has No Advantages over First-Order Logic in Capturing Aspects of Ordinary Mathematical Proof

That, in the examples of 7.5, the predicates and relations are the linchpins for the neat way "symmetrical" proofs relate to each other (so that mathematicians can allude to the symmetry of proofs without further ado) may seem to advantage *second*-order logic over the first-order sort. For by ascending to explicit (second-order) quantification over such relations and predicates, one can (in some cases) capture the relevant symmetries axiomatically; and thus what appears to be something mathematicians just "see" about how proofs are related translates (in the second-order context) into explicit (although topic-specific) conditions on *relations* and *predicates* that can be axiomatized.

---

[26] Consider as well the proof that the derivative of a function at a local maximum is zero, and contrast that proof with the one that the derivative of a function at a local minimum is also zero.

I plan to dampen your enthusiasm for the thought that the second-order approach has real advantages over the first-order approach. To do so, I first rehearse an objection of George Boolos's (and, following his, Stewart Shapiro's) that the requirement that mathematicians (as a practical matter) be capable of surveying or otherwise getting a psychological grip on proofs they understand and accept, seems to show that if derivations, in some sense, underlie *proof,* those derivations had better be second-order, not first-order, *even in the case in which the inferences involved are first-order ones.* Here's the argument. Boolos (1998a) exhibits an example of a valid argument *I* in a first-order language, and indicates a short (second-order) derivation of *I* from its premises. However, "it is well beyond the bounds of physical possibility that any actual or conceivable creature or device should ever write down all the symbols of a complete derivation in a standard system of *first*-order logic."[27] He concludes, later in the essay (380): "The fact that we so readily recognize the validity of *I* would seem to provide as strong a proof as could be asked for that no standard first-order logical system can be taken to be a satisfactory idealization of the psychological mechanisms or processes...whereby we recognize (first-order!) logical consequences."[28]

Now this claim is misleading, although both Boolos and Shapiro show keen awareness of the point I'm about to make: Of course, it isn't just that *any old* first-order logical system fails so miserably to capture how we so readily recognize the validity of *I*; it's that the *specific* first-order system which *isn't* allowed to quantify over the functions and relations (that the second-order proof—in its way—*is* allowed to quantify over) that so miserably fails. That is, a *first-order* system that helps itself to additional objects (functions and sets) over and above the numbers, such that the sentences of the inference *I* is restricted to quantifying over, is one in which a proof *equally short* (nearly enough) is available.[29]

Shapiro (1999, 46), in a polemic directed against Burgess 1993, glosses this as the adoption of first-order *set theory;* in so doing, he assimilates the suggestion to a foundationalist practice, and adds, "[Burgess's] idea is that first-order set theory provides a *uniform* foundation for mathematics, useful for many purposes" (italics mine).[30] Shapiro then writes, warningly, "The background set theory has a staggering ontology." And he concludes, "Burgess

---

[27] Boolos 1998a, 376.

[28] See Shapiro 1999, 46, where this sentiment is (more or less) quoted, but fully endorsed. Also see Shapiro 1991, 124–26.

[29] Boolos (1998a) expresses awareness of the point in the paragraph on page 380, which begins: "It may be remarked in passing ..." Note also (i) that I've just now also responded to the worry this section opened with: First-order systems with richer resources can make explicit the facts relied on about predicates and relations; and (ii) that, in any case, the DI view *doesn't* require that the first-order derivations indicated be (psychologically) surveyable.

[30] Burgess (1993, 364–65) may have something different in mind. He writes: "Little of the work in X-theory consists of deducing theorems from the X-axioms.... Results are proved about X-structures 'from the outside', and such work is perhaps well represented as making deductions from first-order set theory." He *adds*: "Of course logicians are aware that the full strength of the usual systems of set theory is not really needed: more restricted systems would in principle suffice."

conceded [!] that logicians are well aware that the full strength of set theory is not really needed."

This is also misleading, but in *three* ways. First, one surely wonders how a second-order system, with "*standard* model-theoretic semantics," and in which "in each interpretation, the property or set variables range over the entire powerset of the domain d, . . . etc.," is supposed to be ontically *austere*—even comparatively.[31] Second, and much more important, imposing ontic strictures on *mathematical* practice is anyhow pointless; mathematicians *never* worry about the purported ontic costs of their practices. That suggests that any philosophical position that implies a *genuine* ontic cost to be had (in *mathematical* practice, when, that is, additional "ontology" is introduced) has badly misconceived that practice.[32]

Third, Burgess's point about mathematicians proving results about deductions in X-theory "from outside" hardly requires a uniform characterization (in full **ZFC**, say), as Burgess notes. The practice can be quite adequately studied within the confines of an approach (e.g., mine) that uses an open-ended family of *first-order* systems and draws upon (explicit and implicit) axioms about additional objects (of various sorts), such as specific relations and/or specific sets of numbers, and so on. That, after all, is how *mathematicians* proceed: They introduce "new objects" and bring the properties of those objects to bear on what's already being studied. That the objects usually are specific relations and properties of items already under study is why set theory is such a good context to "uniformly" study this practice—if that's what one wants to do.[33]

Given, therefore, that second-order logic provides no advantages, at least as far as feasibility of proofs is concerned,[34] over a construal of the practice that takes mathematicians to (routinely) augment (first-order) algorithmic systems by the (ad hoc, but mathematically valuable) introduction of new

---

[31] Shapiro 1999, 42. A *confession*: This objection is ad hominen. Second-order logic with standard semantics, as it's commonly presented (and Shapiro follows suit in this), employs nonempty domains in which massive commitments to pure sets appear in *every* domain. Perhaps (although I haven't seen an explicit presentation of this) these commitments can be finessed from the logic by allowing an empty domain, much as a commitment to *something* can be expunged from first-order logic. In any case, see the second point below.

[32] See Azzouni 1994, especially part I.

[33] I've lapsed (and have done so for a while) into an ordinary "ontic" way of speaking for the nonce. But as we'll see in 7.9, this won't involve any genuine "ontic commitments."

[34] Shapiro (1991, 208) writes: "The previous chapters of this book *assume* working realism [the view that "most of the discourse of informal mathematics can be taken at face value . . . and that it successfully refers to structures that are unique, up to isomorphism. . . . [N]o position is taken (so far) concerning the metaphysical nature of the indicated structures"] and argue from that perspective that first-order languages are inadequate models of mathematical practice. . . . If the natural language of mathematics is cast in a first-order language, then one cannot account for the characterization and communication of the presumed structures." I've previously described exactly this as "the problem of referential access," but Shapiro (1999, 58) strangely disapproves of such language. Avoiding it, we can still say: Concern with categoricity is restricted (historically, and today) to (certain) logicians and set-theorists; *categoricity* is simply not an issue for the ordinary working mathematician—not now and not historically. Why? Because the working mathematician presumes successful access to such objects if he or she has definitions or axioms

objects, are there any disadvantages? Well, yes, *if our aim is to capture salient facts about mathematical practice.*

Recall 6.3 and 7.3. One point made in those sections is that although mathematicians have their disagreements (just as other kinds of practitioners do), in a sense those disagreements don't cut as deep. Mathematicians—and now I'm excepting twentieth-century developments—disagreed over which concepts and methodologies should be introduced into their practice, but they didn't disagree over the *implications* that mathematical statements have. That is, there is a sense in which they tended not to disagree over proofs, as they would have if the relevant steps in a proof were ones to be verified by an ineffective *semantic* implication relation. Second-order logic, with standard semantics, blurs the sharp distinction between the concepts (and objects) introduced into a mathematical subject matter and the implications that such introductions have. But if mathematical practice reflects a sharp distinction on this point (and I think the historical record shows it *does*), then any codification of mathematical practice that eviscerates our (algorithmic) grip on the implications of our mathematical claims seriously distorts that practice.

7.7 More Details on the Interplay of Mathematical Proof and Algorithmic Systems

The most common objection to views which take formal rules (algorithmic systems) to be the *modus operandi* of mathematics is that mathematicians don't have these rules in their heads.[35] This objection misfires if raised against the DI view. For, first, that position doesn't require mathematicians to be conscious of (all) the rules of the system (they're implicitly working with) any more than it's required of ordinary speakers of natural languages that they be conscious of the rules *they're* using.[36] Second, the picture doesn't require mathematicians, in

---

that (more or less!) distinguish such objects *from the others under study.* The real question always is: Can new concepts be introduced into this context consistently that yield results about such and such objects (i.e., when can we augment systems fruitfully?). This is a question that can be illuminated by set theory (by showing that set theory, as a whole, can be introduced successfully, and that the concepts involved are definable in set-theoretical terms). Categoricity itself provides no advantages whatsoever—and "working realism," if it *really is* ontically neutral, is consistent with this because, presumably, it's consistent with *nominalism.*

I should add that my general views about treating contemporary mathematical practice as operating within the confines of algorithmic systems—and where designated subject matters are treated within open-ended families of such systems (and where new kinds of objects can be introduced into any subject matter provided only that they pay off proof-theoretically) is entirely compatible with the inclusion of second-order algorithmic systems in such families. I'm only intent on showing that there aren't any particular philosophical benefits to so doing.

[35] See, e.g., Kreisel 1967, 153–54.

[36] One must not draw too precise an analogy here. I've avoided calling the view, say, the "derivation-deep-structure" view—borrowing, that is, linguistic terminology—because in linguistics the deep structure of a sentence *A* is one that by a series of transformations results in the surface syntax of *A*; and this isn't the case, or so I've urged, when it comes to proofs and derivations.

any case, when studying a subject matter, to remain within the confines of a single (algorithmic) system—indeed, if anything, mathematicians are required to transcend such systems by embedding them in larger ones. The derivation indicated (by the application of new tools to a given subject matter) can be a derivation of the weaker system the mathematician started with, or it can be a derivation of a stronger system, (some of) the terms of which are taken to pick out the same items supposedly referred to in the weaker system.

I should stress this, however. It doesn't much matter exactly where in the family of algorithmic systems we take "the" derivation indicated by a proof to be located. The reason for this is that the transliteration of the ordinary proof into its derivational proxy relies on making explicit tacit assumptions in the proof—and there is intrinsic vagueness in precisely what's tacitly relied on in an ordinary mathematical proof. It's a reasonable, but quite rough, lower bound on the derivation that all the explicit concepts of the ordinary proof are proxied in it. (The view would be that if a derivation exists in which many of those concepts were missing, it would illustrate that many of the steps in the ordinary proof were ones that were—unknown to the designers of the proof— redundant.) The concepts explicitly in play may not be sufficient for a derivation in many cases because of the tacit assumptions also required. Some of these may be items that mathematicians were clearly aware of—but many needn't be. The resulting derivation, of course, doesn't have to be practically surveyable; but this isn't a problem since mathematicians aren't understood as—psychologically speaking—*grasping* such derivations (when they survey the ordinary proof correlated with such derivations). Similarly, there is a rough but reasonable upper bound: One is loath to include concepts in the derivation clearly alien to the proof—as it appears in ordinary mathematics, anyway—unless the derivation is impossible without them. These rough guides allow a lot of latitude as it's often unclear whether certain concepts appear in a proof tacitly or not: Since the algorithmic systems—embedded in one another—preserve the result of the systems they augment, it's hard to see what problems this indeterminacy can lead to.

It's also worth adding a second reason for the studied ambiguity in the location of the derivation indicated: This is that a proof often focuses as much on the notation used in the proof as it does on the purported objects talked about. For example, a mathematician, in the course of a proof, may note that a certain indexing system is well grounded, and so that a certain inductive process thus terminates. When formalized as a derivation, such involves meta- mathematical elements that drive it into a derivation in a system strictly larger than one about, say, the objects officially under study. Mathematicians auto- matically ascend to a discussion of what can be taken to be properties and

---

Also, leaving aside (syntactic) ambiguity, each sentence corresponds to an item it's the result of (by transformation). But I argue momentarily that there is a much more pervasive systematic ambiguity about which derivation a proof can be taken to indicate (and in which algorithmic system). Because of the fact that algorithmic systems are cumulative in their results, this doesn't matter.

relations of the relations and properties of the objects they're proving results about; and they don't particularly worry about whether in doing so they've augmented the mathematical system they're implicitly working within (introduced, that is, "new objects" that bear a hierarchical relation to the ones under study—and that implicitly augment the proof-theoretic properties of the system they were heretofore implicitly working within).

One might still be worried: The *derivations*, as they arise in different algorithmic systems, aren't the same: They are (generally) shorter in stronger systems, and they involve different terminology. Yes, but why is the mathematician required to have (even subconsciously) a *particular* derivation in mind? Why doesn't an indifference about exactly which derivation is being indicated by a proof suffice for mathematical practice? It's not, after all, that the derivation(s) indicated are (somehow) the "logical form" of a proof in ordinary mathematics; it's that some derivation (or other) is indicated, and that hardly requires the uniqueness of the said derivation. (See 7.11 for a discussion of the worry that we can't even claim that the *theorem derived*—as it appears in different algorithmic systems—is the same one.)

7.8 How the Practice of Proof (but Not Derivation) Presupposes Knowledgeable Readers

Since the day-to-day practice of mathematics isn't to actually *execute* derivations, but only to *indicate* derivations to themselves or to others in their profession, it's clear why *proof* and not *derivation* occupies center stage in mathematical practice; and this despite it being, in a very clear sense, *derivations* which provides the skeleton for (the flesh of) *proof*. As we saw in 7.5, the topic specificity of the mathematical subject matter arises in a crucial way here: How algorithmic systems are augmented turns on the *nonlogical truths* already in place.

But there is another way, more sociological in flavor, in how topic specificity intrudes. This is in how mathematicians presuppose *knowledgeable readers*: Certain proof procedures are assumed as part of the toolkit; a certain level of competence (in proof) is assumed as well, so that if a proof takes a certain form, the reader is taken as able to modify it appropriately. These presumptions, turning on the background knowledge (and ability) of the profession at a time, can't be codified—not once and for all, anyway. This provides one reason for textbooks in mathematics, and why they become dated.

Here are a few examples. The mathematician is assumed to have readily available a number of familiar (and yet flexible) techniques—such as induction—that, with very little ingenuity (given the right training), can be applied to the case at hand. "Familiar," as already indicated, is largely a matter of education, and so it not only changes over time, but also is relative to an intended audience (mathematicians in a specific field; students at one level rather than another, etc.). One then writes, for example, "Emulating the proof of Theorem 2.9.1, we can easily prove . . .," or "What we did for A we can also do for B," or "We first do it for two groups—not that two is sacrosanct. However, with this experience behind us, we shall be able to handle the case of any finite number easily and with dispatch . . . ," or "All these verifications have a certain sameness to them, so we pick one axiom . . . and prove it holds," or "Repeat the argument on this relation with π_2. After n steps, the left side becomes 1, ..."[37]

Related to this, it may be that one proof resembles another not in any neat specifiable way but in many small ways, so that by changing this and that and that and that (and a few other things, as a result of the earlier changes, as one goes through the structure of the proof) one sees how to get the new result. That is, there is a cascade of changes in the proof structurally resulting from a couple of initial changes: And yet although the entire set of changes may be elementary, they can elude a simple syntactic characterization that could be encapsulated in a subsidiary logical inference. That is, what has to be changed in a proof can be quite open-ended from a syntactic (derivationally theoretic) perspective, and yet is something trivial for a mathematician. Here we find phrases such as: "Clearly what we did for 3 we could emulate for any integer n, in which case the factor group should suggest a relation to the integers mod n under addition," or "We modify the proof given . . . ," or " . . . and indicate the slight change in argument for . . . ," or "The proof does not depend on the characteristic of F and holds equally well even in . . . ."[38]

Moreover, there is a hodgepodge of cases in which it's assumed that, given the knowledge (and skill) of the intended audience, certain details, proofs, and so on aren't needed either, because (i) (in a textbook) the reader should learn the material by supplying the details him- or herself, or because (ii) the professional can easily supply what's missing, or because (iii) what's missing is too tedious to supply and the professional can just *see* that it's true, or because (iv) even if the professional can't be relied on to just see that it's true, the professional can rely on the author having (successfully) determined it's true. (E.g., "By a simple induction, the corollary extends . . . ,"[39] or the ubiquitous "It is easy to prove," "It is easy to see," or "It is an easy matter," or "expanding this out and making many uses of Lemmas 6.9.2 and 6.9.3, we obtain. . . ."[40])

MacKenzie (2001, 321) writes of ordinary mathematical proof ("rigorous-argument proofs of ordinary mathematics") that

> [a] sense of audience is crucial to this form of proving: a sense of what listeners or readers will know, and of what will be enlightening; of what the audience will understand, and what they will not; of what needs to be spelt out, and what can be covered by phrases such as "it is obvious that . . . ," or "we can show similarly that . . . " or "without loss of generality, we can assume that . . . "; and so on.

MacKenzie's characterization of the social factors that affect when something is admissible among mathematicians as a proof, and consequently,

---

[37] Herstein 1975, 72, 105, 104, 134, 148, respectively.
[38] Herstein 1975, 52–53, 204, 324, 327, respectively.
[39] Herstein 1975, 164.
[40] Herstein 1975, 328.

why what's acceptable as a proof can change over time, is clearly compatible with the DI view.

7.9 The Ontic Burden of Mathematical Reference

I must discharge a promissory note regarding a kind of ontic *façon de parler* that I've freely indulged in during the course of this chapter. I've described mathematicians as introducing new "objects" into old contexts in order to prove results about the original subject matter. It should already be clear (although still worth making explicit) that the ontic-free translation of this is to note that one algorithmic system has been embedded in another, and in such a way that terms in the more impoverished system are treated as co-referential with (some) terms appearing in the terminologically richer system. Co-referentiality, in the mathematical context, has only a logical role that doesn't involve objects: What's allowed, when terms are taken to co-refer, is substitutivity *salva veritate* in all extensional contexts.[41] The *new objects* brought into the picture (invented, that is, by the wily mathematician) are ghostly reflections of the new terminology that have been linked to the old terminology in certain mathematically valuable ways (e.g., we can now also "refer" to relations taken to hold among the "objects" previously "referred to"). And these two observations, quote marks and all, exhaust the purported "onticity" of mathematical reference.

The absence of relata for mathematical terms doesn't deprive them of "content" with important roles. It's natural, perhaps, to assume that if there is no object for a term to refer to—that anchors the role of that term—then it can't have a role in discourse distinguishable from other terms without referents. But this is as false of mathematics as it is of fiction. In the latter case, although (many) fictional terms fail to correspond to anything that exists in any sense at all, they nevertheless have rich and distinguishable roles. Whether such roles relate to the "semantics" of such terms is an issue I'll table for now. In any case, the content—I'll continue to use this term—of the sentences of ordinary mathematical proofs is similar in its abilities to that of the sentences about purely fictional objects in this respect: Neither is "meaningless," but neither— on the nominalist view I favor—describes anything that exists either.

Two ways in which the "object-centered" appearance of mathematical talk plays important roles are worth pointing out. One way, which I won't dwell much on here (although I have elsewhere), is that such "objects" often code (some of) their empirical applications. Geometric notions, *point, line*, and so on, do this nicely. Indeed, our tendency to treat these notions as idealizations of physical items, extended locations and borders, is better understood as code for the nature of the approximations needed to facilitate these applications. More pertinent to the focus on proof is the second way, that the "objects" referred to by mathematical terms can act as guides to fruitful

---

[41] And, in mathematics, this is *all* contexts, nearly enough.

ways of augmenting algorithmic systems—introducing fresh axioms, that is. This can happen, for example, if a mathematician codes certain proof-theoretic properties of a set of objects as itself a kind of object with specifiable properties—thus the naturalness with which mathematicians move from descriptions of the properties and relations of objects to the study of those properties and relations themselves. Often, as we've seen, this correlates to ascending metamathematically from one algorithmic system to another of greater strength.

It should be said: To speak freely—as a nominalist—of *there being* entities that *exist in no sense at all* isn't doublespeak: "There is" is an indispensable verbal device for speaking of "things" (e.g., the contents of dreams or hallucinations) regardless of one's ontic attitude toward such. Philosophers have long known that "there is" and, indeed, "there exists" *needn't be* ontically committing:[42] The native speaker uses such locutions without having a paraphrase within reach, and without thinking ontic commitment is afoot, as talk of average men, possible ways, houses one's planning to build, and so on, makes clear.[43]

7.10 Tacit Assumptions

The sentences in an ordinary mathematical proof seem to have content that isn't about derivations, or indeed, about proofs (of any sort) at all; such sentences seem straightforwardly about *objects*—abstracta. And that raises a question about how the noncommitting content of mathematical statements relates to the indication of derivations. How does "indicating" come into what ordinary mathematical sentences are doing (or seem to be doing)?[44]

No ordinary mathematical proof indicates a derivation in the sense that it (or a designating term contained within it) *refers* to such a thing or describes its properties. Rather, as we've seen (7.5 and 7.8), an ordinary mathematical proof is a combination (often a complex one) of having a particular form, coupled with explicit allusions to suppressed details, that *shows* that a derivation (of such and such a sort) exists. Consider, as a first illustration of the phenomenon, an ordinary mathematical proof **P** that is (say) an abbreviation

---

[42] Quine, despite his endorsement of his triviality thesis regarding the ontic role of the first-order objectual existential quantifier, the claim that we learn the meaning of the quantifier via the ordinary expression "there is," nonetheless reveals a sharp awareness of this. He (1980a, 107) writes: "In a loose way we often can speak of ontological presuppositions at the level of ordinary language, but this makes sense just in so far as we have in mind some likeliest, most obvious way of schematizing the discourse in question along quantificational lines. It is here that the 'there is' of ordinary English lends its services as a fallible guide—an all too fallible one if we pursue it purely as philologists, unmindful of the readiest routes of logical schematization."

[43] See Varzi 2002, for examples, and for citations of other philosophers sensitive to this point about natural language. See my 2004a, part I, for arguments that ordinary uses of "there is" or "exists" can't be paraphrased away, treated nonliterally, nor taken as ontically committing as a matter of their *semantics*.

[44] My thanks to Adam Morton for raising a version of this question.

of a derivation **D**. **P** indicates **D** *not* by containing terms that refer to **D** but by virtue of the fact that anyone familiar with the process of "deabbreviating" proofs knows how to exhibit **D** on the basis of **P**. Although—most of the time—ordinary mathematical proofs aren't abbreviations of derivations in this tight sense, the abbreviation model of indicating is still applicable. If "the argument is symmetric," or if the text presupposes the reader can construct a series of simple inductions, and so on, it's still the case that anyone familiar with the process of "deabbreviating" proofs in this broader sense knows what it takes to exhibit **D**. Notice that this can hold in principle even if the actual construction of the derivation is (because of limits in time and energy) impossible.[45]

But, of course, I've claimed that mathematicians, for most of the history of mathematics, have been unable to "deabbreviate" proof into derivations— that they've not known that this could even be done, or even what it would *mean* to do such a thing. How is that possible? Even granting the empirical evidence of the success of programs to recast ordinary mathematical proofs into derivations so that they are amenable to mechanical recognition procedures, one can still wonder by what mechanism *mathematicians* provided proofs that indicated derivations without having any clue that this was what they were doing. The key to this epistemic puzzle turns on the use of tacit assumptions. It's notorious that mathematicians often (perhaps nearly always) rely on tacit assumptions only teased out by later practitioners;[46] indeed, tacit assumptions are a perennial aspect of mathematics. One way that (substantial) tacit assumptions are overlooked is that they are seen as "obvious,"[47] and by this I don't mean that they are consciously seen and marked as "obvious"— rather, it's that in the normal flow of showing *this* on the basis of *that*, the mathematician overlooks them.

How is this possible? Two reasons: First, mathematicians often gain access to proofs through idealized models of the subject matter (drawn from implicit applications) that enable them to presume on the tacit assumptions. Tacit geometrical assumptions are often overlooked in precisely this way: One sees at a glance that any simple closed curve on a plane separates it into precisely

---

[45] I'll make more of this later: Note that the mathematician (implicitly) knows how to execute the transformation on the proof—the view isn't that he's got the result of the transformation represented in his mind (even implicitly) already. See 8.2 for reasons for thinking the latter is impossible.

[46] Two well-known historical examples (but, of course there are many others): first, the several substantial assumptions that Hilbert (1971) teased out of Euclid; second, the numerous (presupposed) uses of the axiom of choice on the part of mathematicians (like Borel and Lebesgue) (i) who didn't realize that their results depended on such a strong assumption, and (ii) who (for some) repudiated the axiom when later brought to consciousness of it. See Moore 1982. Post facto "rigor" is often a matter of teasing out implicit steps in a proof. Thus the contrast between the "rigor" of textbook presentations of proofs, and those of the originals—e.g., Riemann—is precisely in the tacit presuppositions made explicit, often as intermediate steps in the proof that (according to the textbook standards of the time) need to be shown.

[47] And, often, what makes them—to later practitioners—no longer "obvious" is that denying them offers alternative approaches.

two components, of which the curve is the boundary. But showing this by direct geometric arguments (the intuitively natural way) is surprisingly quite difficult.[48]

It might seem that this sort of example (which is rampant in mathematics) shows that the content of mathematical statements really does contribute to an appreciation of inference. In one sense it does: The mathematician is unaware of tacit assumptions built into the models he or she uses to understand the subject matter. Nevertheless, such tacit assumptions aren't—when exposed—seen as part of the proof procedures but exactly as I've described them: tacit assumptions. Fresh mathematics arises by seeing what tacit assumptions have been overlooked and seeing how fruitful negating such assumptions turns out.

Notice this crucial aspect of mathematical practice doesn't fault the suggestion that ordinary mathematical proofs correspond to mechanically recognizable derivations: Mathematicians often tease out tacit assumptions, and build new mathematics on their alternatives. The recognition, on the part of the profession, that the assumptions are tacit, shows recognition that the proofs in question are incompletely analyzed—as opposed to a recognition that such proofs are complete, but that the semantic content of mathematical terms is the basis on which the proofs have their validity.

A second but related reason why tacit assumptions are easily overlooked is that mathematicians often devise algorithms for generating theorems, results, or formulas, where tucked into those algorithms—algebraically, as it were—are such tacit overlooked assumptions.[49] Nice examples of this occur in algebra—where one teases out what sorts of properties hold of more general algebraic structures: that, for example, the reason that the integers can be embedded in the rational numbers is that the integers are a commutative ring with no zero-divisors, and the rational numbers are a field, and anything like the former (any "integral domain) can be embedded in a field, the way that the integers are embedded in the rationals. Often the application of an algorithm operates on the basis of concepts that subsequently prove not to be the right ones—mathematically speaking—for analyzing why the algorithm works.

Several things have been shown by the foregoing. The appearance of a subject matter (i) prevented mathematicians from seeing tacit assumptions, and thus (ii) prevented mathematicians from seeing how much further analysis of their proofs could be taken, and thus (iii) prevented mathematicians from seeing that in fact it was the existence of derivations from such tacit and explicit assumptions, the recognition of which mathematicians were agreeing on.

It's worth adding that even when mathematicians realized—as they always did—that steps were missing in their proofs, they could nevertheless overlook *how much* was missing. After all, a tacit assumption—in a sense—is

---

[48] And showing it topologically isn't trivial either!

[49] One often has to laboriously unveil what assumptions are presupposed in a particular algorithm to prove that it can (or can't) apply to new cases.

(psychologically) invisible; otherwise it wouldn't be *tacit*. (To realize that *something* is missing is hardly to realize *how much* is missing.[50])

7.11 Rav's First Issue

I turn now to three issues raised in Rav 1999.[51] The first seems to strike at the heart of the DI view. Rav claims that when faced with the claim that *B* follows from *A*, the process of analyzing the intermediate steps that take us from *A* to *B* "has no theoretical upper bound. In other words, how far has one to analyze a claim of the form 'from property *A*, *B* follows' before assenting to it depends on the agent" (14–15). Rav cites Kreisel's (1970) invocation of Gödel's theorem as, if "properly interpreted," supporting this insight.

What's going on? Why should anyone think that a *finitary* piece of mathematical reasoning—say, a step in a proof—corresponds to something infinitary (if, that is, we attempt to translate it into a derivation)? A clue (apart from the reference to Gödel's theorem) is afforded by Rav's citation of Steinbring 1991, where we read that "knowledge must be interpreted . . . as a complex structure which cannot be extended in a linear or deductive way, but rather requires a continuous, qualitative change in all the concepts of a theory" (15 n. 8).

To elucidate this, first consider this (perhaps naïve) view of *concepts*. The *content* of a concept, let's say, is to be captured by a set of axioms which exhaust all the truths about that concept.[52] If this notion of content is taken to apply to mathematical concepts, they look overrich and open-ended as a result. Consider the concept of a counting number; we know there is no way to capture (axiomatically) all the content of this concept—where "content" is so understood.

So what? Why, that is, haven't I just repeated, in the jargon of "concepts," a point shown by Gödel's theorem, that axiomatizations of sufficiently rich subject matters are impossible? Well I *have*; but the thought is that in

---

[50] This helps explain why it isn't a requirement of the DI view that mathematicians, generally, be aware that their proofs indicate derivations. What *is* required is that there is a theory of mathematical practice—that includes the theory of algorithms plus Church's thesis—and that implies, in each case of an ordinary mathematical proof, and based on the properties of that proof (the way that the form of a deabbreviation of a proof is based on the latter's properties), that a (family of) derivations exists. I discuss this further in 8.5.

It also unveils one mechanism in the "tacitness" of principles that allows there to be no fact of the matter (among a narrow set of alternatives) about exactly *which* principles are involved. One can recognize that something can be done without realizing there are variants in how it can be done. Also, different people may (naturally) think of different ways to do something, should they try to be explicit about it.

[51] There are several arguments in Rav 1999 that I won't discuss explicitly. Some of these, for example those in support of second-order logic, I take as already adequately countered; others, for example his positive suggestions about the role of logic as "cohesive," I regard much more favorably but think *can't* explain the datum, in his own words, that "no phlogiston-like case histories [in mathematics are] known" (Rav 1999, 29).

[52] Recollect note 27 from 2.7.

giving a derivation from (a proxy of) A to (a proxy of) B to underwrite the step (in the mathematical proof) from A to B, we have a problem we won't see *unless* we couch the Gödelian insight in conceptual talk: that the mathematical *concepts* involved in A and B are ones whose "content" isn't captured by locating a purported indicated derivation in a *particular* algorithmic system. Crudely put, the *concept* of number (for example) can't be captured by any (effective) axiom system. And this means that the step from A to B is open to a potentially infinite analysis because the concepts involved in A and B are open to such an analysis—as we ascend to richer and still richer axiom systems, say.

We've seen an innocuous version of this already: I mentioned in 7.3 that mathematicians routinely supplement a subject matter with new concepts, and in so doing facilitate proofs about old objects. But what's being stressed here is that, from the strictly algorithmic-system point of view, it isn't only that new concepts are added to the mix, it's also that old concepts are *augmented*—they gain additional *content*. Indeed, if one is strict about concept-individuation conditions, what can be claimed is that the new systems come with all-new *concepts*—and the old ones have simply been stipulatively identified with (some of these) new concepts. Such a stipulative identification of concepts— that proves valuable—is innocuous solely because of the cumulative way that algorithmic systems are embedded in one another: None of the old results about the old set of concepts is jettisoned—new material has only been *added*.

From the perspective of ordinary mathematical practice, however, the picture looks different. We are discovering new things about our *old* concepts; so (it must be that) our mathematical concepts really *are* richly open-ended; and our ability to grasp such concepts in mathematical practice allows us to generate an open-ended set of results about those concepts that can't be connected to any particular algorithmic system; furthermore, when analyzing a step in a proof, there is always more to say, because there is always more to say about the concepts that licensed the inference.[53]

However: Just as it's sensible for us to deny apparent commitments to mathematical objects that we're supposed to somehow (magically) reach out to from a family of algorithmic systems, so too it's sensible for us to deny that we continually pull out new content from the infinitely deep concepts that we've been gifted with, as opposed to the (humanly available) picture that we just continually augment our concepts by consistently augmenting the algorithmic systems that such concepts arise from. There's no harm, I hasten to add, in loosening identity conditions on concepts (whatever *those* conditions *are*) so that we can remain systematically ambiguous about what algorithmic system (in particular) we're working within—provided only that we don't allow this way of talking to *mystify* ("*denaturalize*," perhaps, is the right

---

[53] These two views of mathematical concepts correspond closely to the contrast between "broad" and "narrow" content as this has played out in philosophy of mind. The contrast turns on the individuation conditions of concepts: whether or not they involve (at least partially) the objects that the concepts are about.

word) what grasping mathematical concepts is. In order to see what's really happening, we must recognize the convenience of continuing to speak of the same concepts even when—strictly speaking—the rules governing those concepts change as we shift systems.

Perhaps I've been unfair to the view Rav has alluded to: Instead of thinking of concepts as having an endless fund of content all on their own, what's involved is a "holist" view of the content of concepts: One determines the content of a concept by means of contrasts to other concepts; and it's these contrasts which provide a endless fund of content to a concept.[54] I've no particular disagreement with the view—so stated—and I'm willing to claim that since the concepts in question arise from augmentations of derivation systems, that the DI view itself construes concepts holistically in the sense so described (see the next paragraph). But this holism doesn't lead to the conclusion that an infinite or open-ended analysis of the role of a concept *in a particular informal proof* looms—informal proofs are fairly strict about which contrasts (and other relations) are or aren't relevant to the use of particular concepts in them, and this is the case even when a proof is reconstrued in a stronger algorithmic system.

Recall two reasons from 7.7 for the systematic ambiguity in where a derivation—indicated by a proof—is located in a family of algorithmic systems, or, more accurately, exactly *which* derivation is indicated by a proof. A third reason for this (already mentioned in this section) is that despite augmenting algorithmic systems as routinely as they do, mathematicians take themselves as exploring the same (central) set of concepts even when a change in the context (caused, say, by the new tools introduced) leads to an ability to prove new results about those concepts. If one (like a good *logician*) notices when changes in algorithmic systems occur, one is aware that new axioms are in play, and that these operate synoptically with the old ones so that one can't always neatly explain what the new derivational power of the system is from; in a trivial sense it's the new axioms, of course; but (of course) those new axioms *all on their own* don't enable those derivations. But if one thinks like a (Platonist) *mathematician*, then one thinks not in terms of shifting the axiom system one is working within (say) but about the introduction of new objects into an old setting to illuminate familiar objects; and then one's intuitions about how *objects* are individuated naturally takes the old objects to be unchanged by the introduction of these new objects (our being able to prove new things about those old objects is glossed *epistemically*, as it were: Our new tools enable us to prove things we couldn't prove before about these old objects). This is one way in which the "object-centered" perspective (in contrast to the view focused on algorithmic systems, or more specifically, axiom systems) makes it natural to think that introducing new concepts in an old area doesn't stop our saying that doing so illuminates concepts "we already had about objects we were already studying."

---

[54] My thanks to the referee (for the paper this chapter is drawn from) for raising this concern about my fairness to Rav's suggestion.

The reader may be worried. If our concepts *really* change when we (implicitly) shift on the algorithmic system we're working within, then there is a sense in which theorems from different systems can never be genuinely identified with each other (since the terms appearing in them aren't—"intensionally" speaking—the same). Doesn't this create problems for the perspective that the derivation of a theorem indicated by a proof is ambiguously located in a family of algorithmic systems? No (and here I'm repeating a point made earlier in this section), we (implicitly) stipulate a co-referential identification among *terms* in different systems; this in turn induces an identification among statements in different systems, and of course, an identification among concepts. When mathematicians accept a proof of a theorem, they have recognized a derivation (of that theorem) as located somewhere in a family of algorithmic systems; and so it's located in every algorithmic system that algorithmic system is embedded within—regardless of what new concepts subsequently come into the picture or how our previous concepts become "enriched." So just as mathematical concepts seem strangely and endlessly "deep"—so too (and for the same reason) do ordinary mathematical proofs.

7.12 Rav's Second Issue

Many highly developed mathematical theories (as they are practiced) are unaxiomatized; and (on the face of it) this looks problematical for the ordinary formalist, although not for the DI proponent. Rav (1999, 16) gives several examples: (i) matrix theory, (ii) graph theory and combinatorics, (iii) probability theory, (iv) number theory, (v) group theory. Each illustrates the remark made in the opening paragraph of 7.5 that mathematical areas of study are rarely (if ever) restricted to particular algorithmic systems; the benefits of augmenting such systems with new terminology are too valuable to resist. In some cases (Rav's (iv) and (v)) one sees that a piece of the theory *has* been axiomatized (sometimes *famously* so), but routine mathematical practice brings in material from anywhere else (provided it leads to valuable results); the augmenting of the particular (first-order) language of PA, or of group theory, is naturally open-ended. In the other cases, the different algorithmic systems that are tied together—by the co-referentiality of (some) of their terms—to study the intended subject matter are so open-ended and so rich as a group that axiomatization of even an interesting part remains elusive.

Why? Well, it's rare that a *single* axiom system proves interesting enough, or illuminating enough (mathematically speaking), to drive a mathematician to explore it on its own. Many subject matters—and this is crucial—*are in the normal course of their development* best described as involving a continual ascension from one algorithmic system to another as new "concepts" are (continually) introduced. The whole process (in classical fields, anyway) can be embedded in a portion of ZFC—that is, one *particular* algorithmic system—by a long (and no doubt tedious) sequence of *definitions*. Carrying this out explicitly would (no doubt) illustrate graphically why axiomatizations of these (entire) fields haven't (really) been attempted.

I should add that Rav's inclusion of graph theory indicates what I've pointed out already: that algorithmic systems can be diagrammatic in nature. They needn't be restricted to sentential form—as long, of course, as the derivations (in diagrammatic form) are mechanically recognizable (in principle).

7.13 Rav's Third Issue

The last point Rav (1999) raises that I'd like to say something (more) about is the topic specificity of mathematical proof. He stresses that typical moves in a proof "bring to light the intensional components in a proof: they have no independent logical justification other than serving the purpose of constructing bridges between the initially given data, or between some intermediate steps, and subsequent parts of the argument. But the bridges are conceptual, not deductive in the sense of logic" (26).

It's important that the target, even if it's the strict formalist that Rav usually has in mind, and not the DI proponent I favor, not be unfairly characterized. It's true that (some) formalists thought that formalizing mathematical proof would bring epistemic benefits; but surely we can separate the formalist who thinks such benefits include the mechanizing of proof production from those who think it includes only the mechanizing of proof recognition. The latter position is compatible with undecidability results; the former isn't. Some of what Rav writes about the topic-specific nature of proof suggests that he takes his target to think that proof construction (and not just proof recognition) can be mechanized.[55] Since the former is possible in principle only when a decision procedure is in hand, and rarely possible in practice even then, we should leave this particular issue aside.[56]

Still at issue, perhaps, are two points: first, that mathematicians rely on intuitions of some sort to determine what should be introduced into a subject matter to yield results, and second, that what's introduced is usually not deduced from what was there already. The second point is one that treating mathematical subject areas as *families* of algorithmic systems explains: What amount to new axioms (and terms governed by them, that aren't already given in the subject matter the proof is ostensibly about) are introduced routinely by

---

[55] I'm speaking here especially of the discussion to be found on pages 26–27 in which Rav stresses how "topic-specific" moves are regularly introduced in proofs and that "no rule ever suggests which is the appropriate move to be made in a given situation."

[56] I'm also leaving aside Rav's points about the fact that the theorem proved is often not as valuable as the methods introduced to prove that theorem. On one construal of the point, it's also agreed to by all parties in dispute; technically speaking, a formalist considers *every* step in a proof to be a theorem proven, not just the last item in a sequence that happens to be written down. Of course, "methods" may be more "abstract" aspects of a proof—*ways* of constructing proofs, or even informal methods for indicating proofs. Only the last is something the strict formalist hasn't access to.

mathematicians in the course of a proof—and in this way the algorithmic system the mathematician is implicitly working within shifts. The first point, alas, is something *all* of us would like to know more about: how great mathematicians come up with the right concepts to ply to make the (previously) intractable tractable. It's true that imagery of various sorts has a role in telling the mathematician what can be compatibly asserted about a kind of object: In this sense, mental models can be used to provide a guide to how an algorithmic system is to be augmented. Here again an analogy with novelists may help: Novelists, when making up characters, often think about real people; they add real traits to the portraits they're constructing to see if the resulting combinations will work. So too a mathematician might think (implicitly) of various applications, either empirical ones or (more likely) applications to other mathematics, and use such to attribute properties or relations to the objects under study. *This* aspect of mathematical creativity isn't quite as *forced* as theorem proving is—an observation illustrated by Lakatos 1976: We can define things this way *or* that way, and not every way of doing so is equally of mathematical value; but this is far from showing there is only one way to go (and this is why a kind of object can lead to *many* different sorts of mathematical studies, e.g., subsets of the real line).

But there doesn't seem to be anything about the operation of *ordinary* mathematical proof that reveals anything special about this process: The creative mathematician is good at supplementing algorithmic systems to lead somewhere, and the rest of us aren't. If there were something special about ordinary mathematical proof that indicated how to do this, we would all learn how to invent new techniques of proof by studying such proofs instead of learning from them, as we do, only how to *imitate* them—sometimes creatively—in new contexts.

7.14 Concluding Remarks

On many (hostile) construals of formalism, mathematicians are taken to be caricatured by that view as deduction-drudges—engaged in the mechanical computation of mathematical results. On the DI view, the mathematician is seen instead as gracefully sprinting up and down algorithmic systems, many of which he or she invents on the spot. The contrast in viewpoint isn't merely that the latter has a prettier picture of the mathematician's job than the former, although there is that (and it isn't negligible); the point is that there are two issues we must negotiate successfully if mathematical practice is to make philosophical sense. The first is that, as we've seen, mathematics—as a knowledge-gathering practice—really is different from other ways of procuring knowledge. It's not that in mathematics we can't get things wrong—as philosophers have too often suggested—it's that mathematicians have a way of agreeing about proof that's virtually unique (agreement among humans is always a surprise and always has to be explained somehow), and the explanation of this surely is that some sort of recognition of mechanical procedures is involved. Against this appealing move is that should we actually try to *replace* (informal) mathematical proof with a specific set of derivation procedures that *really are* mechanically recognizable, and that seem to determine the same class of theorems as ordinary proof in ordinary mathematics, we get something no mathematician wants (and rightly so). I've tried to show wherein mathematical practice such mechanically recognizable derivations live (not on the surface of that practice), and why, nevertheless, from their secret hiding places they ruthlessly stamp mathematics with their unique mark.

In doing so, I've naturally worried about two foils to my position. The first is some form of Platonism, a view that tries to explain the agreement among mathematicians in terms of their grasping the referents of mathematical terms. Empirical science also seems successful at inducing agreement among its practitioners. The story of this is a long one,[57] but ultimately the explanation turns on epistemic processes that generate relations *to* the objects studied in science—relations we can exploit for information about those objects. Jettison Platonism and any similar relation to abstracta, and one fears that content-external sociological factors are all that remain available to explain agreement among mathematicians.

I've shown that this dichotomy is a false one: The nominalist who denies that mathematical terms refer to anything at all needn't, as a result, sign on to Strong Programme sociology. This does leave a promissory note, however: One wants a story about the algorithmic systems I've apparently committed myself to. Are *they* represented in the minds of mathematicians, or are they abstracta which mathematicians grasp?[58] This question, too, introduces a false dichotomy, as I show in chapter 8.

---

[57] My version of it may be found in Azzouni 2000b.

[58] This latter position returns us to Platonism, through the back door as it were: Abstracta arise not as relata of mathematical terms—as what mathematical terms refer to—but as the material that derivations are ultimately composed of. An abstracta-dependent version of formalism thus survives, although nominalism does not.

8

How to Nominalize
Formalism

8.1 Introduction

Nominalists and formalists, when instantiated in real flesh and blood, are sympathetic with one another: Both groups, whether real or not, hate taking mathematical vocabulary at metaphysical face value—acquiescing in the existence of abstracta, that is, and letting such abstracta and their properties dictate the truth conditions of mathematical statements. But thereafter, their paths fork: "Discourse-criterion nominalists," obsessed with ontology (and, of course, Quine's criterion for what a discourse is committed to), worry about applied mathematics, and either engage in rewriting (applied) mathematical text to remove quantifier commitments to abstracta, or else steep themselves in the tangled reanalysis of applied mathematical doctrine to show that scientific explanations and/or the justification (confirmation) of scientific doctrine don't—contrary to appearances—turn on presupposing the truth of that doctrine. Traditional formalists, meanwhile, focus on pure mathematics, and hope to show the in-principle replacement of mathematical proofs by derivations; that is, they hope to show the in-principle replacement of ordinary proofs in ordinary mathematics by sequences of sentences that take mechanically ("effectively") recognizable forms.[1] Chapter 7 revealed (apart from my nominalism,

---

[1] Except where it matters to the argument, I leave tacit the relativity of derivations to algorithmic systems. I should add that although nominalists, of the sort I've alluded to, actually *are* instantiated as flesh and blood philosophers, "traditional" formalists (nowadays, anyway) only seem to subsist thinly as caricatured foils in the literature.

which I'd previously admitted to), a commitment to a *kind* of formalism: It's the (tacit) coordination of mathematical proof with derivations, that one or another derivation is indicated by an ordinary mathematical proof, that explains why mathematical practice has the form it has, and has (sociologically speaking) the properties it has.

My hope is that this explanation doesn't require mathematical abstracta that different mathematicians grasp (and whose grasping explains their agreement). For then Platonism is powerfully *reanimated*: not by what mathematical terms refer to, but by what derivations themselves are *made up of*. But if one presupposes derivations-as-abstracta, and allows the mathematician's mental manipulation of them to explain mathematical practice, why balk at abstracta as references for mathematical terms, even despite the accompanying (mystically flavored) story of "grasping" their properties to explain agreement?

Any view which takes derivations to be crucial to ordinary mathematical proof must grapple with the fact that derivations, so understood, aren't on the surface of mathematical practice; ordinary mathematical proofs, far from being mechanically recognizable sequences of sentences, involve, as we've seen, a rich topic-specific interplay of concepts with (Arbib 1990, 55) "somewhat large jump[s] from statement to statement based on formal technique and on intuitions about the subject matter at hand." The point of this chapter is to win a perspective on how derivations can play the role the DI view gives them without thereby introducing an ontic commitment to derivations as a species of *abstracta*.

8.2 Objections to the Neurophysiological Representation View of Derivations (NRD)

Let's sharpen the issue: If derivations do play the role in mathematical practice required by the DI view, where exactly are such role-fulfilling derivations *located*?[2] If they aren't themselves mathematical abstracta, then the only other option (it might seem) is that they are representations in the minds/brains of mathematicians. This would connect derivations (rather directly) to the understanding that mathematicians have of ordinary proofs: One or another derivation would psychologically underlie (and enable the comprehension of) an ordinary proof. But how plausible is it that actual derivations—in all their detail—reside (somewhere) in mathematicians' minds? To explore this question, compare this "neurophysiological representation view of strict derivations" (hereafter, NRD) with the Chomskian view of the syntactic processing needed to recognize and produce grammatical sentences of natural languages. Although such processes aren't ones ordinary (or even extraordinary) people are conscious of, such processing is still "psychologically real" in the sense that

---

[2] I owe this blunt way of posing the problem to David Isles (personal communication); my thanks also to Adam Morton for raising a version of the same problem.

exactly which transformations the "language organ" facilitates, and what general principles it uses, are open to empirical discovery.[3]

In the case of mathematics, it's not of course that one must posit a "mathematics organ" in the sense that I-languages are posited by Chomsky as the expression of a "language organ."[4] Nevertheless, still plausible is an empirically less ambitious position wherein *some* innate mental capacities result in neurophysiologically embodied sets of axioms and derivation rules (that generate derivations). These rules can differ from mathematician to mathematician, and can differ over time for a single mathematician, depending on (i) the particular branch of mathematics being studied (howsoever such branches of mathematics are individuated by the cognitive architecture of said mathematician), and (ii) how that mathematician augments over time the rules he or she applies to the subject matter. Nevertheless, such rules, according to NRD, underlie—in a way analogous to how the rules incorporated in the empirically posited language faculty issue in an I-language—the surface texts of proofs that mathematicians actually write down.[5]

Here are some objections to NRD that might occur to readers. First, it may be urged, derivations of *B* from *A* are often far too long to be items that, psychologically speaking, mathematicians actually process *to recognize* that *B* follows from *A*. Note the point: Chomskian views about the language organ are compatible with its containing rules (finitely stated ones) that imply that natural language sentences—*of whatever length*—are grammatical. But such views *don't* require that the sentences that speakers are *able to recognize as grammatical* presuppose arbitrarily long mental representations. To require *that* would disbar such representations from being "psychologically real." However, when (particular) mathematical proofs that mathematicians do (routinely) comprehend are replaced by derivations, the latter are often—perhaps

---

[3] See, e.g., Chomsky 1986, 7–13, 25, and esp. 36–37, for the form (circa 1986) such empirical studies then took, and that he expected them to take in the future.

[4] Dehaene (1997) notes that many psychologists, however, posit an "accumulator" by which animals and humans recognize small numbers of objects, a suggestion that has nicely amassed large amounts of empirical evidence. This accumulator, although yielding a topic-neutral concept of number (one tied to no particular sensory modality), operates *approximately*. Apart from this, the internalized sense of number intuitively available to adults is also limited—and misleading—both in the mathematical concepts it exemplifies and in the properties of those concepts it registers. Dehaene writes (40): "In essence, the number sense that we inherit from our evolutionary history plays the role of a germ favoring the emergence of more advanced mathematical abilities." I should add that the empirical study of psychological *dispositions* that support mathematical competence results in what should have been expected: Such dispositions are limited in their scope and range, and have inherent weaknesses such that they can't explain our grasp of current culturally mature mathematical concepts.

[5] Dehaene (1997, 6) writes, "When our brain is confronted with a task for which it was not prepared by evolution, such as multiplying two digits, it recruits a vast network of cerebral areas whose initial functions are quite different, but which may, together, reach the desired goal." This leaves it open as an empirical possibility that different mathematicians neurophysiologically instantiate the same subject matter differently; that in fact one reason why creative mathematics is so hard is because the mathematician's brain must solve the neurophysiological task of theorem proving (for certain kinds of mathematics) *innovatively*.

usually—implausibly too *long* to be suitable candidates for psychologically real representations that enable mathematicians to understand the original proofs *and* that they are proofs. For representations to play that role, *they* would have to be amenable to psychologically real transformations that connect *them* to the proofs in question.

Second: Even if the tedious computation of a derivation *is* neurophysiologically processed by a mathematician's brain (without her knowing this) while that selfsame mathematician works out a proof, there is another problem. As already indicated, the mathematician's recognition that *B* follows from *A* must be tied to the underlying derivation that (according to NRD) is neurophysiologically realized by the mathematician's brain. But what derivations actually look like makes it impossible for such things to underlie *anyone's* comprehension of a proof. The texture of details—ones utterly irrelevant from the mathematician's point of view—is overwhelming. How is the mathematician's mind to take such things—which on this view have to (literally) appear in that mind (somewhere) to be manipulated—and abstract from them the higher-order patterns that enable her to write down an ordinary proof?

A third problem is that, as we've seen (recall 7.8), many of the ways ordinary mathematical proofs deviate from derivations aren't local to the proofs; such proofs, that is, aren't abbreviations of derivations—even when "abbreviation" is broadly construed.

If we keep in mind the open-ended nature of the algorithmic systems supposedly instantiated in the brains of mathematicians, these objections can, to some extent, be mitigated. For example, derivations look *terribly long* when couched in a system such as Russell and Whitehead's PM or Quine's ML; for then every result must be traced back to an initial set of premises. Derivations of particular results from such "first principles" rapidly become unwieldy.[6] But in practice, mathematicians don't prove things from "first principles": They reach for whatever (nearby) results have been shown to follow from a system; convinced that **P** is so, either by eyeballing the proofs themselves, or on authority, they can then establish new results on that basis. In effect, they operate (at a time) within an algorithmic system that takes **P** (among other things) as a premise. We can similarly allow the derivations taken to psychologically underwrite their results to contain derived rules and other shortcuts based on specific theorems that hold of the subject matter, as long as the resulting algorithmic system has a mechanical recognition procedure for proofs. These considerations respond not only to the first objection, but also to the third.

It may be felt that such derivations are still too long, for two linked reasons deriving from the second consideration given above: first, the simple fact that ordinary mathematical proofs themselves—which are a far cry from anything mechanically recognizable—are often already quite long! And second, that because even the foreshortened derivations, contemplated as a response to the first and third considerations, must be mechanically recognizable, this usually

---

[6] See Azzouni 1994, part III, § 2, where this point is used to argue for a derivational division of labor among mathematicians.

requires enormous detail. Cases in ordinary mathematics in which mechanical procedures are explicit, not in theorem proving but in computation, already reveal how daunting the many details for an entirely explicit mechanical procedure must be. Work in programming reveals the same fact. I'm assuming, of course, that the algorithmic systems contemplated as psychologically real don't help themselves to results that the mathematicians writing or comprehending the ordinary mathematical proof don't have access to. For otherwise a desirable algorithmic system is *easy* to find: Let **A** be the (finite) conjunction of premises used by the mathematician, and let **B** be the desired result, and let **A** plus **A** $\Rightarrow$ **B** be among the premises of the algorithmic system (plus modus ponens)!

We might try to respond to the second consideration this way: Those partial to the Chomskian program repeatedly note that rules executed by the language organ—and indeed, the computational (and other) processes needed for cognition of all sorts—needn't be consciously accessible.[7] Indeed, it's typical for the categories that ordinary speakers (consciously) bring to bear on sentences when interpreting them, say, not to be categories that the currently best empirical theory, of how such sentences are processed, implies that speakers are using.[8] This point, however, applies best not to the mental machinations of original mathematicians, but to the flashy but ultimately pedestrian abilities of calculational wunderkinder. It would be no surprise if such (often autistic) wunderkinder manage their astonishingly rapid calculations by quite specific neurophysiological tasking.[9]

I resist extending this suggestion to creative mathematicians because derivations are (more or less) fixed in their vocabulary, but great and original mathematics introduces *new* concepts not definable in the algorithmic system then under study—and often not belonging to any algorithmic system already known of. Even if we think the mathematical mind has an infinite Fodorian stock of such concepts representable within it (that can be conveniently introduced into proofs when needed), some (neurophysiological) story must still be told about how such concepts are constructed from the mental storehouse of primitive concepts in an orderly yet profitable manner—the computational task looks overwhelming if derivations must always be deployed to test the fertility of introducing one or another concept.

---

[7] See the methodological homily in Fodor 1975, 52–53, esp. n. 19.

[8] Fodor (1975, 49–50) on sensory mechanisms *in general*: "There is . . . quite good empirical evidence that an early representation of a speech signal must specify its formant relations. Yet speaker/hearers have no conscious access to formant structure and, for that matter, very little conscious access to any other acoustic property of speech. It is, in fact, very probably a general truth that, of the various redescriptions of the input that underlie perceptual analyses, the degree of conscious accessibility of a representation is pretty well predicted by the abstractness of its relation to what the sensors specify."

[9] Dehaene (1997, 7) gives reasons to avoid even *this* concession: "Like the rest of us, experts in arithmetic have to struggle with long calculations and abstruse mathematical concepts. If they succeed, it is only because they devote a considerable time to this topic and eventually invent well-tuned algorithms and clever shortcuts that any of us could learn if we tried and that are carefully devised to take advantage of our brain's assets and get round its limits. What is special about them is their disproportionate and relentless passion for numbers and mathematics."

These considerations perhaps don't definitively refute NRD or vindicate it; indeed, the sorts of considerations I've been raising, both pro and con, *can't* vindicate the view in any case because NRD ultimately must be empirically established. It's not enough, that is, to show that derivations from algorithmic systems *can* (in principle) psychologically underlie the practice of ordinary mathematical proof; it must be shown that they *do*. But this bit of methodological caution seems to conflict with a different methodological consideration that (some) philosophers have long found strangely appealing: a kind of "transcendental deduction" of specific theoretical structure on "the only game in town" grounds. If the reasons I've given in chapters 6 and 7 are definitive in explaining why derivations *must* play a role in mathematical practice if certain historical and sociological facts about that mathematics are to be explicable, then (for lack of alternatives) we have what amounts to a transcendental deduction of NRD; more prosaically, we have empirical evidence from the nature of ordinary mathematical practice that neurophysiological representations of derivations *must be* involved.

This may be put in a way that has a methodological dark side: The *explicability* of historical and sociological facts about mathematical practice is beholden to empirical results about neurophysiology. But really (it may be argued), we've seen this before, and it isn't empirically crazy: At one time, evidence about the age of the fossil record implied that the classical theory of atomic structure had to be wrong; this kind of "downward" constraint on science is one that Chomsky has stressed for many years. And, of course, I've been alluding to Jerry A. Fodor's (1975) sort of argument that certain tasks that humans (and animals) are capable of require neurophysiological representations of various sorts. Still one may (and, I would claim—but can't argue for now—should) be worried. It's one thing to have downward evidence that a certain theory is false; it's quite another to engage in a kind of reasoning that establishes a certain sort of theory *along with substantial ontic commitments* on the sheer grounds that—as far as we can tell—there are no other alternatives.

There is, however, another way to preserve a role for derivations in ordinary mathematical proof but to reject the methodologically suspicious and perhaps neurophysiologically implausible NRD. This is the DI view of chapter 7: A standard mathematical proof *indicates* any of a family of derivations without those derivations (i) being what standard proofs abbreviate, (ii) being, in some more extended sense, the "logical forms" of such proofs, or (iii) being items that such proofs are "reducible to." Instead, ordinary mathematical proof, by (among other things) unsystematic combinations of genuine derivation sketches, allusions to such sketches elsewhere in the literature, and metaderivational considerations, convinces mathematicians that the proof is valid because (although the mathematician needn't know this) the proof corresponds to a derivation of such and such a sort.[10]

---

[10] In a sense—but this is an *analogy*—the mathematician engages in numerous local versions of completeness proofs as we saw in chapter 7: that such and such a topic-specific implication (relative to a certain model—a certain semantic interpretation of the terms used) corresponds

8.3 Theories of Mathematical Proof

Some remarks should be made about a crucial assumption of the theory of algorithmic systems, that the derivations (in an algorithmic system) admit of an effective *recognition procedure*: I've already claimed that it is now, and always was, the indicating of such derivations (belonging to one or another implicit algorithmic system) that made mathematical proof epistemically special—in a word, *convincing*. This fact about mathematical proof has remained historically unchanged (it's *supra*historical, if you will). But what did change over time—and *drastically*—were (philosophical) views about what it was that *made* mathematical proof epistemically special; philosophical—more strictly speaking, *epistemological*—doctrines *about* mathematical proof have changed. Descartes' (1931, 19) explanation, for example, was that proofs (with care) are amenable into formulations where "no steps are missing."

Here is a paragraph offering some (provisional) notes on some stages in the evolution of the epistemology of mathematical proof (historians, no doubt, can—and will—say much more about this than *I* can). Plato saw how convincing mathematical proof is, but he didn't pay much attention to *how* proof manages this. He preferred to generalize perception to "Forms." Descartes, Pascal, and especially Leibniz, *did* focus on the mechanics of proof: They (collectively) developed the idea that the epistemic power of mathematical proof is ascribable to its (implicit) basis in primitive concepts from which deduction proceeds without *gaps* in reasoning, and therefore, without a place wherein mistakes can intrude. When the notion of mechanical execution is introduced (by Turing and company), this isn't the realization of that earlier programmatic dream but a *replacement* of that semantic conception of reasoning by a syntactic one. The earlier no-gaps view tried to explain what's so striking about mathematical proof—but it placed too heavy a burden on mathematical practice: It made practitioners of mathematical proof potentially *infallible*.

As I've noted before, "mechanically recognizable" is less constraining than the presuppositions of traditional mathematical proof, since the latter include substantial logical principles. Because these logical principles weren't visible until Frege (and others) *exposed them*, when mathematicians did think about what proof requires, they focused on how convincing proof is, rather than on an innate set of (logical) principles that in part induces this sensation of veracity; this, in turn, allowed twentieth-century alternative logics to be seen not as something new—something *not* mathematics but (at best) only analogous to (real) mathematics—but instead as new developments *in* mathematics. Had the logical presuppositions of traditional mathematical proof been explicit, they would instead have been understood as individuation conditions on what we *call* mathematics.

---

to a mechanically recognizable derivation ambiguously occurring within a family of algorithmic systems.

This is why I adopt the general algorithmic system as the basis for derivations rather than (say) first-order axiomatic systems. Perhaps this goes too far—recall Gaifman's challenge from 6.7, note 34: Perhaps (human) mathematicians can't operate within general algorithmic systems, but only within something classical or (at most) a tame fragment of such (e.g., intuitionist logic). If so, this changes almost nothing claimed *here* about the role of derivations in ordinary mathematical proof, and so those with doubts about my generalization to algorithmic systems *tout court* may take the more conservative position.

Church's thesis is that "mechanically recognizable" is mathematically specifiable (without residue) via Turing machines or via mathematical formulations equivalent to the Turing machine formalism;[11] this is, of course, an *empirical assumption* about which branch of mathematics is applicable to a particular domain of activities that includes (among other things) mathematical proof,[12] an empirical assumption on a par with, say, the choice of one or another framework geometry for the motion of objects through space. Were an algorithm discovered (that humans use) that transcends the Turing machine formalism, we'd change the branch of mathematics (partial recursive functions) applied to this area for another.

8.4 The Ontic Status of Derivations

I discussed, in 7.9, the rich roles of the noun phrases in ordinary mathematical statements despite the nonexistence of what those phrases refer to. But what about the derivations themselves indicated by such proofs; what sort of *ontic presence* must such derivations have on this view? Not much of one. Derivations needn't be instantiated anywhere in space, neither now, earlier, nor later: They needn't be physically real inscriptions in any sense at all, however tenuous. All that's presumed is that mathematicians design ordinary mathematical proofs in accord with the *theory* of algorithmic systems. Despite the lightweight ontic commitments of this theory of algorithmic systems, *now that we know* that mathematical practices accord with this theory, we must be committed to its *truth*.

I've just stressed the "lightweight" ontic commitments of this theory. Before unpacking the metaphor "lightweight," in the next paragraph, I should first circumvent possible worries: Talk of "instantiations," whether there are any or not, invokes a type/token distinction, and it may seem that introducing *types* of derivations undertakes a commitment to abstracta even if those types aren't instantiated. This may be thought for one of two reasons. On the first, the theory of algorithms—the truth of which we're committed to—may itself

---

[11] See Rogers Jr. 1967, esp. the Basic Result (18–19). A brief discussion of Church's thesis (and its use) appears on pp. 20–21.

[12] See Azzouni 1994, 172–74. Church's thesis provides, strictly speaking, an *upper bound* on the algorithmic procedures humans can execute.

speak of "there being" such *types* of derivations. Even so (as I've argued in Azzouni 2004a with respect to theories of any sort), we needn't be ontically committed to such types, at least not on the mere grounds of such "there is" statements belonging to the theory. But ontic commitment to types may be thought to arise for a second reason: that if two items fall under a predicate they must have a *property* in common. This, however, doesn't imply that there are entities—*properties*—needed to explain how that predicate applies to the two items. Nor, in saying this, need I commit myself to a program of eliminating *talk* of properties (replacing it with talk of similarity). It's possible to take it as a fact—as *objective*—that both *a* and *b* are **P**, without having that imply there is something—some *property*—corresponding to **P** (that exists).[13]

In claiming that the derivations indicated by ordinary mathematical proofs needn't be instantiated in space or time, but only that mathematicians need to use proof methods that indicate such things, I don't mean that derivations *are never* so instantiated. An example of a derivation in the sense meant is a construction (with pen on paper) of a diagram, as in Euclidean geometry. These certainly were (and are) instantiated;[14] but many if not most of the (families) of derivations indicated in contemporary mathematics aren't instantiated.

What relation does instantiation—so understood—bear to existence? As a first approximation, we can describe a derivation as really existing if it appears on paper. This tight constraint allows the existence of very few derivations.[15] We can broaden existence constraints on derivations by allowing those derivations, if any, represented in mathematicians' minds, provided such derivations are *fully* represented (every detail is present). One can go further, as Quine and Goodman (1972, 175) point out, increase the stock of available inscriptions to include "not only those that have colors or sounds contrasting with the surroundings, but all appropriately shaped spatio-temporal regions even though they be indistinguishable from their surroundings in color, sound, texture, etc." They glumly add, however: "But the number and length of inscriptions will still be limited insofar as the spatio-temporal world itself is limited. Consequently we cannot say that in general, given any two inscriptions, there is an inscription long enough to be the concatenation of the two."

This last bit follows only if one insists (as Quine and Goodman seem to) that parts of inscriptions can't overlap: that is, that the concatenation of 'A' with 'B' requires 'B' disjoint from 'A' (and *next* to it in space). If we relax this consideration, then a small square of the universe suffices for any inscription, however long![16] Such overlapping inscriptions *are* unreadable; but Quine and

---

[13] There is certainly more to say about this, but I can't do so now.

[14] Prejudices about "rigor" shouldn't prevent understanding what's being claimed. Construction of a diagram "proving something" is a conventionally specified set of moves recognized to have taken place (as they do take place) *by eye*.

[15] Few derivations from few algorithmic systems exist on paper—except as exercises—for who would do so, and why?

[16] Imagine a pixilated sign which flashes messages one letter at a time. Now imagine that it's simply exposing step by step (by lighting up certain pixels so we can recognize the letters) a long inscription already (simultaneously and overlappingly) located among those pixels.

Goodman deserted *that* concern when they offered inscriptions indistinguishable from their surroundings.

Somehow, despite the citation of venerable ancestors, an air of silliness has intruded into what was supposed to have been a profound (and serious) discussion in ontology. To see *why*, we need only ask what the instantiations of such derivations—that we've cribbed out of material stuff, as it were—*do* for us. If all that's required of a theory of algorithmic systems is that it *have* truthmakers, then (as we've seen) truthmakers are cheaply produced. But the sense that our theory of inscriptions is true *because* it correctly describes the properties of such things (so cribbed) is elusive at best. Rather, to push the construction of truthmakers this far is only to design objects to fit a theory that was supposed (on the contrary) to respect *them*. This is why such inscriptions can play no role in explaining how derivations operate in our mathematical practice; they can play no more role, anyway, than sheer abstracta can.[17]

Thus the truth or falsity of a theory of derivations (algorithmic systems) needn't turn on a class of truthmakers that its truth or falsity is due to. And *this* is in broad agreement with nominalism: The truth and falsity of mathematical statements doesn't turn on the purported referents of mathematical terms nor on the properties such referents have (and that mathematical statements supposedly accord with). Instead, a class of mathematical statements (relative to a family of algorithmic systems) are stipulated in their truth values, and this is compatible with a practice that, over time, enhances a family of algorithmic systems with new members—and thus new characterizations of mathematical statements as true or false.[18]

One difference between pure mathematics, as just sketched, and the theory of algorithms that explains how the practices of the mathematician who writes ordinary proofs indicates derivations, is that the theory of algorithms is—as noted earlier—an *applied* branch of mathematics. Pure mathematics is beholden only to mathematical virtues: interest, ease of provability, the

---

[17] Quine and Goodman (1972, 174) offer only this motivation for their programmatic nominalism: "Why do we refuse to admit the abstract objects that mathematics needs? Fundamentally this refusal is based on a philosophical intuition that cannot be justified by appeal to anything more ultimate. It is fortified, however, by certain *a posteriori* considerations. What seems to be the most natural principle for abstracting classes or properties leads to paradoxes. Escape from the paradoxes can apparently be effected only by recourse to alternative rules whose artificiality and arbitrariness arouse suspicion that we are lost in a world of make-believe."

Compare the theory of derivations (and its truthmakers, so construed) with our theories about gold or electrons, and *their* truthmakers. There, our network of theories not only describe properties that gold and electrons have, but also (at least in principle) explain why our epistemic engineering *to them* is successful. They explain how, that is, those properties enable us to get *them* right; we have theories that explain how we're able to discover the properties that these things have. And when we don't have such theories, we still have a *relation* to these things *that we can exploit epistemically*, and the nature of which we can develop a theory about. Demanding something similar of theories about derivations isn't to slide toward verificationism, since electrons, for example, are theoretical entities on *anybody's* story of them.

[18] For how truth is detached from truthmakers (how truth is detached from ontic commitment), see part I.

introduction of more powerful proof methods which nevertheless (consistently) conserve the results of proofs already in place, and so on. But applied mathematics differs: There is an empirical subject matter that the mathematics is applied *to* and that (more or less—depending on the particular application) the branch of mathematics fits. In the case at hand, it's an ever-enlarging body of mathematical results and methods. How is the theory of algorithmic systems coupled with Church's thesis supposed to fit this? If it isn't by truth-makers, by *existing* derivations that mathematicians produce or that they otherwise have access to in some way, how *is* the suitability of this application of mathematics to the practice of mathematics established?

8.5 Truth and Truth of

Truth is cheap—we can (and do) take any body of doctrine as true if it's indispensable: If we *must* talk our way through it on occasion.[19] "True of" costs more—it does, anyway, if what we take a bit of doctrine to be true *of* are objects that we (anyhow) take to *exist*. The most straightforward application of doctrine to a group of objects is to "put" the objects themselves *in* the domain of discourse of the doctrine—in Quinespeak: among the items the quantifiers of that discourse range over. But nothing requires they be *everything* those quantifiers range over: It may be (and often is) convenient to "pad" the universe of discourse with nonexistent entities that, together with the real items that we apply the doctrine to, make up the domain of discourse for that doctrine. One can't, of course, be irresponsible: Although the resulting theory can say about nonexistent items whatever convenience dictates (that's why they're *there*) there are grave repercussions (e.g., falsification) if the theory makes false claims about existing items.

Such padding of the real with the unreal proves *so* valuable for a fairly humdrum reason: The less you can say (in general) about a group of somethings, the easier it (sometimes) is to say those things you *can* say. This may seem a wild strategy: throwing away information about what's *real*. But it's not if, in doing so, there emerge from the welter of details tractable generalizations that otherwise would have remained undetectable.[20]

A caveat: I've committed myself to a "deflationist" view of truth; one according to which the role of the truth predicate is solely for blind endorsement. One may fear that a more substantial view of truth nestles quietly among my views, one revealed by the "true of" idiom. Not so: Statements are true for different reasons, some because of truthmakers and some not. But, as we saw in 1.8, these different reasons aren't part of the theory of truth and play no role

---

[19] There are qualifications: see chapter 2 of Azzouni 2004a.

[20] The point is neatly illustrated in (first-order) model theory: Although the set of truths that hold of every member of a set of models may be axiomatizable, this (generally) isn't the case for subcollections of that set (e.g., compare the models that PA holds of, and the standard model). See Azzouni 2004a, part II, for examples of the "padding the real with the irreal" strategy in physics.

there. Instead they operate in epistemology, in the story of how we *establish* such truths, and what's required of us *to* establish such truths. The "true of" idiom indicates truthmakers only *if* they exist. The jargon "truthmaker," by the way, also doesn't indicate a "substantial role" for "true"—no more than the "true of" idiom does, and for the same reason; the jargon is, in this respect, misleading.

The foregoing claims naturally give rise to the following worry: In what sense does what mathematicians do fit a theory if most of what that theory is about doesn't exist? But this is *easy* to respond to. We often work with theories *about* entities for long stretches of time, only infrequently bringing the theory so developed to bear *against* those entities. Scientific theories, for example, allow us to develop substantial claims about theoretical entities, even when our only access to those entities is indirect (and infrequent). Even theories about observational entities can be theories brought against those entities infrequently and with respect to very few of them. A homespun example is this: I may plan to rearrange the furniture in my apartment. In considering possibilities, I may visualize different arrangements and predict in each case how difficult navigating about the apartment will be. I may, however, test (i.e., arrange the furniture according to) only one or two of the imagined layouts. My "theory" of the possible layouts of my home furnishings, nevertheless, goes well beyond the particular cases actually instantiated in my apartment. What's striking about how we test both our scientific and our ordinary theories is that what they're about, if they're about anything, plays a role only in very specific circumstances. We ordinarily think of the theory as true (of my furnishings) only if *everything* it says about everything it's about is correct; we believe this even if not everything it's about exists. The point being made now is that the nonexistence of much (or even most) of what a theory is about needn't have much impact on our ability to work with that theory (e.g., deduce "there is" statements from it), and apply that theory (to those items among its commitments that exist). As I've stressed, it doesn't even affect our ability to *test* such a theory. In practice we don't (can't) test a theory by bringing it against *everything* it applies to (and that exists). We select judiciously, when we have a choice, and apply it whenever we can when we don't.[21]

On the DI view, mathematicians have a (monotonically increasing) body of proof procedures that indicate derivations. This body of proof procedures can be taken to yield proofs that indicate derivations, when they exist. But the theory that they so indicate derivations is tested *against actual derivations* only when derivations are *actually* produced. Ontically, therefore, the difference between a theory of derivations and a theory of atoms (say) is this: Although we only infrequently (if ever) interact with individual atoms in ways

---

[21] The reader may not be satisfied with this comparison. After all, in the atomic case (say), the atoms exist whether or not we apply a theory (about them) to specific ones; in the derivation case—so I claim—derivations, or most of them, *don't*. I say a little more about this in the paragraph after next to show the difference doesn't matter as far as confirmation and disconfirmation of theories is concerned: Our methodology is, regardless, identical.

that reveal their properties—we rarely have *thick* access to the *atoms* (as opposed to thick access to massive objects *composed* of such atoms)[22]—we, nonetheless, have good (theoretical) reasons to think the atoms we *haven't* such thick access to *also* exist; but although we similarly infrequently interact with derivations, we have no reason to think *they* exist if not explicitly instantiated. We feel differently about atoms because our theories about matter *require* the existence of the many atoms we don't have thick access to: The tables, cats, and so on, that we take to exist are items we also take as *composed of* atoms. This composition relation is a real one: *In principle* we can decompose a table into atoms that make it up, and study them. There's nothing akin for derivations: That an ordinary proof has the form it has doesn't require that the derivations it (ambiguously) indicates exist (the ordinary mathematical proof isn't "made up of" derivations); all that's required of ordinary proof practices is that an ordinary (successful) proof can't be legitimately constructed if the theory of algorithms implies there is no (ambiguously indicated) corresponding derivation,[23] and that's not enough to require the existence of such derivations. That is, all that's really required is that mathematical proof practices accord with the theory of algorithms; derivations that the latter theory is true *of* needn't exist.

This raises a second issue. How do confirmation and disconfirmation work when most of what the theory is about doesn't exist? There are two developments pertinent to the status of ordinary mathematical proof, and whether such proofs indicate derivations. The first one is that we've come to be pretty

---

[22] I'm alluding here to "thick epistemic access": the forging of instrumental access to specific sets of atoms under specific circumstances, for example, in such a way that their properties are indicated to us by that access exactly as sensory access to massive objects indicates the properties of those objects. See Azzouni 1997, chapter 6 of Azzouni 2004a, and Azzouni 2004c.

[23] Here—in an explicitly ontic context—talk of a theory implying that "there is" a derivation is open to misunderstanding: Recall the brief discussion of this in 8.4. But perhaps a more technical presentation of ontic commitment in terms of quantifiers is valuable: Presuming that the theory of algorithms is first order, an ordinary mathematical proof indicates (ambiguously) a derivation only if such an item is a quantifier commitment of that theory (more accurately, a quantifier commitment to an ordinary mathematical proof with such and such properties implies a quantifier commitment to (a class of) derivations). But this doesn't imply that what the theory is quantifier committed to *exists*. We have: $(\exists x)(x$ is an ordinary mathematical proof with such and such specific properties) $\Rightarrow (\exists x)(x$ is a derivation with such and such specific properties). The antecedent quantifier commitment claim we also take to pick out something that exists because it has been explicitly exhibited by a mathematician. The second is *not* (on these grounds, anyway) something we take to exist because it has *not* been explicitly exhibited by a mathematician (and we've no reason to think it has been so exhibited anywhere else). Pertinent to this is an important distinction the Quinean onticist obliterates, between what I call the "quantifier commitments" of a theory and its "ontic commitments". A theory may have implications of the form: $(\exists x)Px$; we can subscribe to the truth of that theory—in as strong a sense as possible—and to the truth, therefore, of "$(\exists x)Px$"; but that doesn't yield an ontic commitment to Ps. See Azzouni 1998, and chapter 3 of Azzouni 2004a. The cited argument is largely made with respect to theories which are regimented via first-order objectual quantification; for it's with respect to that formalism that the strongest ontic claims are (traditionally) made. But, of course, the argument generalizes: *No* formalism, regardless of the semantics it's coupled with, comes equipped by virtue of its logical structure and its semantics *alone*, with a way of reading ontic commitments off of it; for that, *additional* stipulations are required.

sure we've got examples of derivations (they've been, say, *written down*). This isn't obvious (it's not *transparent* that we've got examples of such) because it's easy—especially when working with a meaningful language-based algorithmic system—to unconsciously take steps that aren't mechanically recognizable. In the twentieth century, especially, we've learned to circumvent this (to go through the *form* of a proof step by step), and we've also developed *machines* that we believe (on empirical grounds) to mechanically execute decision procedures for proofs. The results are gratifying: Given many ordinary mathematical proofs of the form *B* follows from *A*, algorithmic systems have been exhibited, and within them—also exhibited—derivations of (formal proxies of) *B* from (formal proxies of) *A*.[24]

The second, and more important, development is the emergence of the theory of derivations itself. Once the theory of algorithms becomes explicit (at the hands of Gödel, Church, Turing, Kleene, etc.), mathematicians bring indirect evidence upon the question of whether ordinary mathematical proofs really do indicate derivations: Especially relevant are results about the expressive power of algorithmic systems. Such developments supply needed details to show that an algorithmic system exists[25] that contains a derivation corresponding to a certain ordinary mathematical result.

Both sorts of developments are analogous to cases in empirical science, where the objects in question exist: A scientific theory is usually intractable, in the sense that simple applications of it to what it's about, in order to directly confirm it, elude us. Subtle instrumentation and/or the deployment of new applied mathematics is needed for such direct application of the theory; and when such breakthroughs occur, we do apply it just as—in the derivation case—new abilities to actually construct derivations can provide an analogous test of whether (particular) ordinary proofs really indicate such things. Ordinary scientific theories are also confirmed and supported indirectly by how they bear on other theories and how other theories bear on them. This

---

[24] MacKenzie (2001, 322–23) writes: "Many rigorous arguments in ordinary mathematics have been replaced successfully by formal proofs, using automated theorem provers and proof checkers, especially the AUTOMATH and MIZAR systems. . . . What is most remarkable about these many replacements of the rigorous arguments of mathematics with formal, mechanized proofs . . . [is that] it is a conservative process. Applied to programs, hardware designs, and system designs, efforts at formal, mechanical proof frequently find faults and deficiencies that have not been detected by other means. . . . Applied to rigorous arguments within mathematics, however, efforts at mechanized proof nearly always suggest at most the need to remedy matters that a mathematician would regard as basically trivial, such as typographic errors or failures to state the full range of conditions necessary for a theorem to hold." He continues: "Research for this book has been unable to find a case in which the application of mechanized proof threw doubt upon an established mathematical theorem, and only one case in which it showed the need significantly to modify an accepted rigorous-argument proof [a proof of Newton's]."

[25] "Exists" as used here is *not* ontically committing (in contrast with my use of "exhibited" in the last paragraph, in contrast with many of my uses of "*exists*" earlier in this chapter, and in contrast with the my use of "exists" in the next paragraph). "Exists," as just used in the text above, is on a par with the use of "exists" in "To hear some people talk, you would think that all Dickens's working-class characters were comic grotesques; although such characters certainly exist, there are fewer of them than is commonly supposed" (van Inwagen 2000, 245).

corresponds to the second development mentioned above, when high-powered mathematics is applied to ordinary mathematical practices to prove results about their scopes and limits.

What's interesting is that the same scientific methodology applies whether what the theory is about antecedently exists or not, and this shows that the latter ontic issue is decided not by the truth of the theories in question but by *purely* ontic considerations having to do with when and how we bring the theory against the world.

Let's return to ordinary mathematical proof and its relationship to the theory of derivations. Ordinary mathematical proof, understood as in accord with the theory of derivations, was incredibly successful: Mathematical practice, even without (many) examples of such derivations, was still able to indicate which were and weren't quantifier commitments of the theory of algorithmic systems. To put this point another way: It recognized which derivations would and wouldn't exist if we (contrary, perhaps, to physical law) could produce them.[26] And the proof of this is that when we finally were able to (literally) write (some of) these derivations down or have a computer process them, ordinary mathematical proofs proved successful indicators of derivations. This makes a modalized view *very tempting*. One imagines that if mathematical proofs aren't "made true" by derivations that come to exist (or that, timelessly, exist all along), then they're made true by what exists in nearby possible worlds, where there are richer resources for constructing derivations (and so, where constructing derivations ratifies the derivational claims in *our* world—as codified in what ordinary mathematical proofs *our* mathematicians accept).

Resist *these* otherworldly temptations: The desire to wizard derivations into existence—to engage in *magical ontology*—is another version of a hankering for truthmakers (regardless of how helpful they are) that we saw (in 8.4) at work in Quine and Goodman's genial offer of unreadable inscriptions. A true theory (I repeat) doesn't need truthmakers: What suffices is only that should (some) truthmakers subsequently come along, they'll accord with the theory—as we saw happen in the case of ordinary mathematical proofs. What needs explaining is how mathematical practice remains in thrall to a theory of derivations, since the subject matter of such (mostly) doesn't exist. And the answer is easy: The practice isn't beholden to actual exhibited *derivations*, but only to the *theory* that describes such things. And, even better, it doesn't need to be *explicitly* beholden even to the theory; it needs only to be explicitly beholden to proof practices that—regardless of how they evolve—obey such a theory.

This solution may in turn seem to give rise to another puzzle: Why should a theory that describes things that (for the most part) don't exist prove successful later when such things *come into being*? Here too the answer is easy, and has already been given: We *often* formulate theories on the basis of very small samples. If the theory is well made (not, that is, based on parochial properties of the sample), then it will fit subsequent items that it's supposed to apply to. The theory, in *this* case, applies to rule-governed moves in sequential

---

[26] Again, I'm not presuming this is what mathematicians *thought* they were doing.

practices (games, sets of constructions, patterns of inferences), and the property in common that the theory seizes upon (Turing machine computability) is absolutely central to all the cases. Thus the danger of parochialism is avoided.[27]

8.6 Concluding Remarks (on Ontology and Rule Following)

Here are three last worries. First, despite notes 23 and 25, some may still worry about *my* uses of "there is" and "exists." Sometimes, as when I wrote the phrase (above), "if the theory makes false claims about existing items," it was clear that I meant to be expressing an ontic commitment to the items in question—or at least intending to talk about objects we were to presume really existed. On the other hand, when I wrote, "to show an algorithmic system exists," I indicated in a note that I wasn't using "exists" in a way I took as ontically binding. Something similar may be noted with my (ordinary) use of "there is."[28] The idioms, as used in ordinary language, sometimes indicate ontic commitment and sometimes don't. How do we tell which it is? The way we always do: context and subject matter. In particular, I used ordinary devices of style, stress, rhetorical enhancing—in short, the usual literary paraphernalia—to indicate ontic commitment, and to disallow it; and when I feared misunderstanding or wanted to punctuate the point, I dropped a note! As far as ordinary language is concerned—the language this chapter is written in—this is business as usual.[29]

Second, it may be thought that the DI view is yet one more incarnation of that Platonism that *just* won't *stay dead* (despite the *repeated* attempts of philosophers to bury it and stamp *all over* its grave);[30] it could be seen as just one more attempt, that is, to embed something ahistorical and unchanging into the historical flux of mathematical practice. Once upon a time, one directly invoked Platonic objects: That's been rejected. One then attempted to disguise such objects as psychological entities, as representations in the

---

[27] Two caveats. First, the needed repetition is often present in only an idealized way: Below certain thresholds—at certain magnifications (as it were)—genuine repetition may always prove elusive. What's perhaps crucial is an (evolved) capacity to categorize events, that one has trouble distinguishing from one another, as repetitions of the same event; more so, to treat events *as* repetitions of the same event (doing this is crucial to being able to play games). Second, it's always contingent, when exploring a phenomenon in nature—our behavior, for example—if we're going to stumble across central and powerful principles. Whether that happens turns quite simply on the phenomenon: It *happens* that rule following is a *quite* general aspect of our behavior. Perhaps that's because it's a requirement for *intelligent* behavior.

[28] Well, as I went over my uses of "there is," in this chapter, I saw no intended commitments for any of them. (Of course that's a contingent fact about the uses of "there is" that happened to make it into this chapter—certainly "there are As" *does* get used in the vernacular to indicate ontic commitments to As; apart from that, perhaps *there is* something I overlooked.)

[29] See Azzouni 2004a, chapter 5, for further argument that there are no idioms in the vernacular *semantically designated* to indicate ontic commitment.

[30] One fears that philosophy is the *one place*—outside of Hollywood, I mean—where, given enough time, *anything* can rise from the dead.

mathematical mind: That's been rejected too. But there's still a fallback position, actually *two* fallback positions: Don't put the objects themselves—derivations—into the minds of mathematicians, put the *theory* of derivations into the minds of mathematicians; indeed (and here's the second version), don't even put the theory of derivations into the minds of mathematicians, put a practice of mathematics into their minds and make sure that practice is *in accord* with the theory of derivations. The upshot, however, is the same: A purportedly historical yet nevertheless *unchanging* methodology—a methodology that itself seems immune to historical change—is stipulated as present in mathematical practice. But why should there be anything like *that* present and operating in mathematical practice at all? why should any practice of *humans* involve *anything* unchanging?

Philosophers sensitive to the evolving (historical) nature of human practices, even ones—such as mathematics and empirical science—that we're most tempted to characterize as sources of unchanging truths, often seem to think that such sensitivity entitles them to a "flux" metaphysics: a Heraclitean view of *everything* as engaged in constant mutation at all levels and at all times. But what changes over time and what doesn't is purely an empirical matter. The mere fact that something persists in time and takes up space does *not* rule out its immunity to change. Nothing is more paradigmatic of an historical theory than the theory of evolution, and its application to microbiological functions. And yet DNA has proved to be of ancient vintage: The same molecule has played the same role for eons. Even more galling for biological Heracliteans (who attempt to take refuge in the fact that specific DNA structures evolve) is that *specific* genetic structures are not only of ancient vintage (shared, say, by us and flatworms—of *all* things!), but that *some* of these specific genetic structures play *exactly* the same roles in them and us (e.g., in immune function, or in aging).

The point is straightforward: Anything that humans do *can* prove to have suprahistorical elements to it: If these are central enough, then the resulting practice will show a rigidity over time not shared by other practices lacking such elements. Mathematics, I claim, shows precisely this rigidity. The interesting question, therefore, is what the rigidity (what I call "the benign fixation of mathematical practice") is due to. One shouldn't be dismissive of Platonism even if one eventually rejects it. For Platonism is exquisitely sensitive to *this* fact about mathematical practice; and philosophical *sensitivity* should always be respected.

This leads to a third worry: Despite the theory of derivations being about what, for the most part, doesn't exist, the form this theory takes isn't arbitrary because mathematicians grasp the mechanical execution of rules, and that's *the concept* that structures the theory of algorithms that mathematical proof practices accord with. But it's here, isn't it, that social constructivists, those who think that social factors determine how and why mathematicians take proofs to be successful, bring in their most infamous weapon. Solutions to Wittgenstein's rule-following problem seem to suggest that no one—not mathematical prodigies, not anyone else—grasps mechanical rule following on his or her own: It's not *possible* to do something like that.

The "sceptical paradox," as Kripke (1982) calls it, is quite radical in its import: It seems to infirm an individual's ability to make (*any*) predicate apply to what he or she wants it to apply to, and thus to have a theory say what he or she wants that theory to say about those things (the predicate applies to). I've elsewhere[31] addressed the issue of how we "make" our predicates apply to the world (to the extent that they do, and to the extent that we can make them so apply), and the view developed there *isn't* compatible with social constructivist views of how we fix the references of our predicates. However, some may feel that any view (such as the DI view) that places a premium on an ability to grasp derivations by a *proof practice*—without such (purportedly nonexistent) derivations being (consequently) available for socialized training upon—needs to *specifically* address the rule-following paradox, and not simply say something to the effect of: Well, I've got a *general* story about how we enable our noun phrases to pick out things in the world, and "derivation," after all, is a noun phrase like any other. On the contrary, in light of the rule-following paradox, and (the most popular) solutions to it, it may be felt that any invocation of proof practices in accord with the "theory" of derivations only temporarily delays the inevitable collapse of the view into some version of social constructivism.

I don't agree, but I don't now have the time to consider all the wrinkles raised under the rubric of "rule following." I'll say this much: Consider a different worry: How do we tell that a certain amount of time has passed (and not more or less)? One might try to invoke an internal "time sense," but that would be hopeless—not because we lack dispositions for recognizing how much time has passed (in the absence of external evidence) but because such dispositions are worthless. (This is a case where we really don't all "go on in the same way.") Our solution, of course, is to find regular processes, and to fix the passage of time in relation to them. "Regular process," of course, is a defeasible label: We can find that purported examples (e.g., the movement of the sun) can prove—in time—to be slightly irregular. There is a complicated epistemic practice—with a healthy amount of theory and background knowledge—that we employ to get onto measuring time "accurately."

And the practice works. Although there is certainly more to say about *how* it works, that it does so should be evident. Clearly a story of the very same sort is to be told about how we recognize when a rule is being applied "the same way." It, too, is one that can't fall back on dispositions; but it can't fall back on the (arbitrary) fixation of standards on the part of the community either. Because of other issues that have been intricately tied in with this problem—in particular, issues having to do with normativity—the analogy I've just urged may not be convincing, even as a programmatic suggestion.[32]

---

[31] Azzouni 2000b.
[32] That's why there's more to say about this.

CONCLUSION TO PART II

The story of proof, as I've given it, is an odd one insofar as it implies that systematic illusions about the nature of inference in mathematics have been central to that practice. Start with the fact that mathematical proof occurs in the vernacular with the aid of computational tools such as diagrams, beads, and, later, computers. But the vernacular masks the logic—whatever it is—in a very straightforward manner: Inference can't turn on the visible syntactic features of the sentences. Naturally one looks to the subject matter to explain the intuitive force of validity so impressively perceived, and eventually one tries to ground this knowledge of necessity, that inference seems to carry along with itself, in a mythology of abstracta. A long history of (philosophical) troubles starts right then and there because the abstracta solution to one problem— what are we doing when we provide mathematical proofs?—leads straight to another: How exactly are those posited objects grasped by those engaged in designing proofs?

I try to explain what's going on in a different way entirely. But my solution leads to the dramatic fact that we don't know what we're doing when we prove results in mathematics; this is meant not in Russell's infamous sense that we can compute without realizing the import of what we're doing, but in the more drastic sense that what the subject matter seems to be about isn't what's driving our recognition of what follows from what—despite our powerful intuitive impressions to the contrary.[1] This means that our

---

[1] This is worth stressing: Only in certain branches of logic, or in computer programming, I imagine, does one *phenomenally* experience the flavor of designing mechanically recognizable

phenomenal grasp of how we reason is utterly misleading: For it seems to turn on the nature of the subject matter we're reasoning about; indeed, any (ordinary) notion of *understanding*, in the sense that one understands why something follows from something else, seems to clearly require not recourse to the explication of a mechanical process, but recourse to something irreducibly semantic.

This explains why the sort of view of mathematical proof that I attributed to Descartes, Leibniz and others in 8.3 seems intuitively far more satisfying than the algorithmic suggestion on show here. For their idea is that we're working with a set of concepts, not all of which are explicit. But concepts are irreducibly semantic: They're *about* something, and it's what they're about that drives the nature of inferences using them. In this way, the Platonist picture can offer another snare to those already inclined toward concepts; for Platonism seems to provide relata for our mathematical concepts, relata that we can subsequently individuate our concepts in terms of. Thus, "counting number" isn't to be individuated by the principles we hold as governing it, but rather, by the kind of object the concept picks out. This allows us to change the principles governing the concept without realizing that this is what we're doing. It also raises, in another form, the "grasping problem" of Platonism: How is it that such concepts succeed in picking out sets of objects when we can't lay out (once and for all) the principles that govern such objects?

In a way, the story I've been telling here in part II is one that both the later Wittgenstein—with his repeated warnings about the snares of language—and the later Nietzsche—with his repeated observations about how wrong we usually are about ourselves, how wrong we invariably are about the most *intimate* facts about ourselves—might find attractive. For the suggestion is that we're wrong about something we take to be quite fundamental about ourselves: How we reason. This *isn't* a story that destroys the significance of reasoning; it doesn't suggest that reason is incoherent or deeply misleading in some way. Rather, it's simply that we're wrong about what we think we're doing when we reason. Perhaps the point is better put this way: All the explicit psychological indications of what's happening when we draw a conclusion B from an assumption A aren't where the (inferential) action is.

In part III, I turn to a deepening of the story of proof, by seeing how a version of it plays out in the semantic tradition started by Tarski.

One last caveat. Some readers may be puzzled—given my (admittedly tentative) rejection of NRD in 8.2—that I worked so hard to address the Boolos and Shapiro argument in 7.6 to the effect that the psychological processing of inference couldn't be first-order. First, as I just noted, NRD hasn't been definitively rejected because it's an empirical claim. (Nevertheless, I think it's false.) But second, the real issue of contention is my rejection of the move behind the

---

proofs. Otherwise one's practice feels utterly semantic: This, more than anything I think, explains why the mathematician is naturally a Platonist.

Boolos/Shapiro argument: that our understanding of first-order inference requires restricting such inferences to particular (and relatively austere) systems.
Rather, the picture, as it emerged in the later sections, is that inference—when
not explicitly confined to a particular algorithmic system (e.g., reasoning in
ordinary mathematics)—is never so restricted.

III

SEMANTICS AND THE NOTION OF CONSEQUENCE

To gain exhaustive acquaintance with [reason itself and its pure thinking] I need not seek far beyond myself, because it is in myself that I encounter them, and common logic already also gives me an example of how the simple acts of reason may be fully and systematically enumerated.

Immanuel Kant (1998, 102)

Nothing [in the inventory of all we possess through pure reason] can escape us, because what reason brings forth entirely out of itself cannot be hidden, but is brought to light by reason itself as soon as reason's common principle has been discovered.

Immanuel Kant (1998, 104)

INTRODUCTION TO PART III

Part I ended with the peculiar thesis that not only is ordinary language inconsistent, but—despite its inconsistency—it's rational for most practitioners of the vernacular to believe otherwise, even when faced with evidence to the contrary, i.e., liar paradoxes. What lies behind this odd result is that what's rational for an agent to believe is relative to the evidence he or she has conscious access to, or more accurately ("ought implies can"), what evidence he or she can (reasonably) be expected to have conscious access to—given the urgency of the case and resources of time and energy. So, as with other "mental teasers," ordinary reasoners recognize that logical puzzles can—despite appearances—involve fallacies in subtle ways: that one's inability to solve such a puzzle (find the flaw in a fishy-looking proof with the final step: $0 = 1$) doesn't mean that a deep and irresolvable inconsistency has been unearthed (as opposed to some easily rejected presupposition overlooked).

The inaccessibility of aspects of inference to—even sophisticated—introspection continued as a theme in part II, where a sustained argument was developed that the logic at work in mathematical proof was unknown to its practitioners until sometime in the twentieth century. Its inaccessibility to introspection follows immediately from this fact alone.

But it isn't as if mathematicians (and other reasoners as well) were without any introspective indicators for successful inference: Clearly, everyone *seemed* to perceive the marks of logical consequence—even if they weren't always clear about the extent of logic's domain: where, that is, something followed from something else by sheer logic alone, and where there were unarticulated assumptions presupposed (which made the matter not one of pure logic alone). In any case, one seemed to have access to a notion of validity, and it was the validity of an inference that one seemed to recognize. It has been fairly widely presumed that this notion of validity was given precise formulation by Tarski.

When alternative logics first began to be studied, they were approached axiomatically, or proof-theoretically. One system of logic had been laid out plainly for all to see, and so logicians could now vary the rules to see what alternatives looked like. But once Tarskian model-theoretic tools came into their own—the semantic construals of logical systems: the characterization of a class of validities by the construction of a class of models in which they all hold—those tools supplanted axiomatics, so much so that it came to be perceived as a relatively minor matter whether the validities in a particular model theory were axiomatizable or not.[1]

In part, purely mathematical reasons motivated the shift in interest among logicians. Model-theoretic approaches to logic involve rich portions of mathematics: much more set theory, for example, than ordinary mathematics does, as well as a great deal of ordinary mathematics itself—abstract algebra, topology, and so on. Proof-theoretic studies, by contrast, are rather austere. But this is hardly the whole of it. A much deeper reason for the shift—I speculate—is the wedding of the notion of validity, as it's explored in the model-theoretic tradition, with our intuitive notion of validity. The rules that have emerged as paradigmatic of logic—even in their (most natural) natural deduction form—are hardly what we introspect (in practice) when we see that *B* follows (logically) from *A*.

Whether the model-theoretic construal of the ordinary notion of validity really does capture that notion has been challenged—most notably by Etchemendy (1990)—and I have a bit to say about his views in part III. But my ultimate concern is slightly different. It's to explore the ordinary (semantic) notion of validity in order to see (i) whether the uses of *truth* and *modality* that appear in it can be construed in accord with the deflationist view of truth laid out in part I, and to see (ii) whether the "syntactic" view of mathematical proof given in part II is compatible with that ordinary semantic notion—as it's given to us phenomenally when we recognize that a proof is valid.

---

[1] In Barwise 1985 (7), we read: "The completeness problem ties up with the first-order thesis and an even older view of logic, where it was seen as the study of axioms and rules of inference. Of the logics studied here, some have a completeness theorem, some don't. If one thinks of logic as limited to the study of axioms and rules of inference, then logics without an abstract completeness theorem will not seem part of logic. But if you think of logic as the mathematician in the street, then the logic in a given concept is what it is, and if there is no set of rules which generate all the valid sentences, well, that is just a fact about the complexity of the concept that has to be lived with. It is this latter point of view that is implicit in the study of model-theoretic logics." I discuss the presuppositions of this passage in 9.1. For now, note that Barwise's version of the "first-order thesis" is, as he puts it (5), that "logic is first-order logic, so that anything that cannot be defined in first-order logic is outside the domain of logic," and that the "abstract completeness" of a set of valid sentences is (7) the recursive enumerability of that set.

Although part III contains only one chapter, it brings together the material of parts I and II because—but not just because—the notion of truth is so central to our understanding of ordinary consequence, and because, nevertheless, the notion of ordinary consequence seems most explicit, and most explicitly at work, in the context of the mathematical proof. The test case, as it were, both of the sorts of truth deflationisms I'm committed to and the derivation-indicator view of mathematical proof, arises therefore in their compatibility with the notion of ordinary consequence. A story must be told of what's involved in that notion—as we perceive it—that's compatible with the doctrines of parts I and II.

This is the goal of chapter 9. It starts by revisiting the question of whether first-order logic should be taken to be the logic of classical mathematical proof (in what I called—in chapter 6—mature mathematics). After making the case that it should, I then turn to the ordinary notion of consequence, and attempt to say (i) what it is, (ii) how it's compatible with the derivation indicator view of mathematical proof, and (iii) what its relationship is with the standard model-theoretic approach to semantics as it has developed post-Tarski.

Chapter 9 is a distant descendent of a talk I gave at a workshop at the University of North Carolina at Chapel Hill, April 16–18, 2004. My thanks to those who heard (and commented upon) that talk; especial thanks go to Otávio Bueno, Thomas Hofweber, Jonathan Kastin, Arnold Koslow, Michael D. Resnik, Stewart Shapiro, and Edward N. Zalta.

9

Semantics and the Notion of Consequence

9.1 Revisiting the First-Order Thesis: Normative and Descriptive Versions

Although, in 7.3 and 7.4, I was tempted by the suggestion that the implicit logic of mature mathematics is first-order, I eventually conceded that, given the evidence then offered, it was acceptable to presume only some fixed (effective) set of principles that sufficed for classical mathematics as it existed at the turn of the twentieth century. Before taking up the ordinary notion of implication, I want to reexamine this issue to see what can be said on behalf of first-order logic. To start, it should be said that the debate—between proponents and opponents of first-order logic that Barwise (1985) alludes to (in n. 1 of the introduction to part III)—is in large measure a normative one: It's a debate about which logic, if any logic, *should be* crowned LOGIC. This is what Quine (1970) was concerned with when he evaluated the technical merits of various alternative logics against the first-order predicate calculus. Such normative concerns loom equally large when Barwise (1985, 5–6)—in opposition to "the first-order thesis," and indeed, in opposition to the idea that *any* logic (in particular) should be crowned our LOGIC—writes:

> Paging through any modern mathematics book, one comes across concept after concept that cannot be expressed in first-order logic. Concepts from set-theory (like *infinite set, countable set*), from analysis (like *set of measure* 0 or *having the Baire property*), from topology (like *open set* and *continuous function*, and from probability theory (like *random variable* and *having probability greater than some real number r*), are central notions in mathematics which, on the

mathematician-in-the-street view, have their own logic. Yet none of them fit within the domain of first-order logic.

We've seen this concern before (in 7.6; see, esp., n. 34). These important mathematical notions aren't *definable* within first-order logic—but expressively more powerful logics exist within which such notions *are* definable.[1] This is what's meant by Barwise's suggestion (n. 1 of the introduction to part III) that the first-order thesis includes the notion that "anything that cannot be defined in first-order logic is outside the domain of logic."

Sheer definability, however, is a highly misleading virtue from the mathematical point of view. The problem is that definability doesn't gain us anything (mathematically speaking) if the apparatus that the notions in question are definable within is proof-theoretically intractable. Since theorem proving is, ultimately, the paramount mathematical virtue,[2] one can't forget that definability *only when in thrall to enhanced provability* matters to the mathematician in the street.

Treating the definability of concepts as a virtue leads directly to treating the categoricity of theories as a virtue as well. A theory is categorical if all its models are isomorphic to one another—if all the models have the same structure. Notoriously, first-order theories fail to be categorical: In particular, PA, the "intended" model of which is the counting numbers, $\mathbb{N}$, has models of arbitrary cardinality (as every first-order theory with an infinite model does). The ability of second-order logic (given the standard interpretation) to provide a categorical theory for $\mathbb{N}$ enables the view that one can define there the notion of countable set—a notion that, Barwise suggests, eludes first-order logic. But, despite this, categoricity remains an artificial virtue because the categoricity of a theory has more to do with what logical (interpreted) vocabulary is available than with any strengthening of what can be shown on the basis of categorical singling out. Corcoran (1980) gives examples of categorical characterizations of $\mathbb{N}$ from which even obvious truths (such as that 0 is not the successor of any number) can't be proven.[3]

---

[1] Barwise and Feferman 1985 is a rich and detailed survey of the literature of those logics as of that date—indeed, precisely that date: Much of Makowsky 1985—the last chapter of the book—is a presentation of "yet unpublished (and unpolished) versions" (749) of papers of Shelah.

[2] *The* paramount mathematical virtue? What about enhanced understanding? Well, in the mathematical context, definability often *doesn't* enhance understanding. A nice example is afforded by most of the model-theoretic logics studied in Barwise and Feferman 1985: The coding of set-theoretic, probabilistic, and topological notions into, e.g., the quantificational structure of the logics, although yielding definability of the notions in question, relies on the antecedent understanding of such notions in their ordinary mathematical contexts. This isn't to fault the evident mathematical value of the couching of such notions in a radically different formalism. As working mathematicians know, doing so can gain accessibility to results not otherwise easily reached, and enhanced understanding sometimes accompanies this process. Barwise (1985, 12–13) offers precisely this as a possible fallout value for the study of model-theoretic logics. But in general the traditional philosophical view that definition is for the purpose of understanding is at cross-purposes with the uses of definability in mathematics.

[3] The article also gives a nice—although brief—history of the emergence of the notion of categoricity, and the slow distinguishing of it from the axiomatizing of a subject matter. It seems

Given this artificial character of categoricity, in particular, the contrast of its artificial quality with that, say, of recursive enumerability—a property of a theory that's robustly independent of any interpretation of the sentences of that theory—it's a bit odd that Barwise, Shapiro, and others, in the face of the weak link between categoricity of theories and their recursive enumerability ("abstract completeness"), presume without argument that the mathematician (in the street) would prize categoricity over enhanced provability. Of course, there is an intuitive presumption that if one can "grasp" an object with a unique characterization, then one has a grip on it that must be informative. But this is clearly false, especially in the mathematical setting, where an ability to prove results about objects trumps any mere unique characterization of them (up to isomorphism). And, as the picture of mathematical subject matters as families of algorithmic systems indicates, "partial grasping"—when continually augmented with more powerful tools—does just fine.[4] It does fine enough to fault Barwise's tendentious suggestion that the first-order thesis includes the claim that notions not definable in the first-order context are "outside" the domain of logic. In a trivial sense—that such notions elude strict definition in the first-order formalism—this is so; but if what's meant is that such exclusion from the domain of logic is a result of such notions' eluding capture *altogether* in first-order terms (in the sense that there are *truths* about them that can't be found through first-order means), then what's meant is quite clearly false.

Of course, none of this particularly favors first-order logic over competitors in the normative debate; although categoricity may not be a virtue when contrasted with enhanced provability, it is—all by itself—hardly a vice. Thus one of the Lindström's (1969) results, that—in Flum's (1985, 77) words—first-order logic "is a maximal logic satisfying the completeness theorem and the Löwenheim-Skolem theorem," hardly shows first-order logic to be a maximally complete logic in any sense that privileges it over competitors, unless one thinks the inexpressibility results implicitly available via the Löwenheim-Skolem property are virtues all on their own. And indeed, there are many ways to augment the expressive resources of first-order logic—by certain generalized quantifiers ("There are uncountably many . . ." etc.), and by other tools, and yet retain abstract completeness.[5]

---

clear from this history of the early postulate theorists that the value of pursuing categoricity was confounded with the (then unarticulated) virtue of recursive enumerability. Corcoran (1980, 205) pointedly does *not* draw the conclusion that "they were misguided in using categoricity as an index of worth of an axiomatization."

[4] A implicit Platonism can give the desire for categoricity some urgency: If we augment first-order systems in the wrong direction, we might accidentally—this is the fear—rule out the objects we really want to grasp (end up including only nonstandard models). But this surely is an illusory worry—especially given that categorical theories, in second-order languages, say, don't offer any (genuine) palliatives because the ineffectiveness of their semantic consequence relation allows (for exactly the same reasons) movement in the "wrong" direction.

[5] Many examples of model-theoretic logics that are expressively richer than first-order logic but are nevertheless complete are to be found in Barwise and Feferman 1985.

Contemporary mathematics—I've urged—is the study of algorithmic systems of all sorts. There is nothing in that characterization that requires a restriction to first-order logic. Matters change, however, if we consider the question of which logic, if any, was the stable underpinning for mature mathematics, and, arguably, is the best candidate for an innate logic—if there is one at all. Here the first-order (descriptive) thesis, I think, carries the day. Let me give the reasons.

First, although the generalized quantifier apparatus seems to naturally occur in ordinary languages, and therefore, many of the ways first-order logic can be augmented to develop model-theoretic logics, such as $\mathscr{L}(Q_1)$,[6] can be easily and naturally expressed in English once certain terminology has been introduced into that language (e.g., by the ordinary English phrase, "There are uncountably many ..."), there is little reason to think such post-Cantorian notions are operative in the logic of mature mathematics, or earlier. Indeed, the very late emergence of a clear notion of infinity—at the hands of Cantor— seems to suggest that the inability of first-order logic to define finiteness is no problem for the descriptive first-order thesis.

There are, therefore,[7] two alternatives to the first-order thesis as a descriptive thesis of the logic of mature mathematics (and earlier): The first is, compatibly with the historical record of a fixed canon of proof procedures, that some (incomplete) axiomatization of higher-order logic is implicitly at work. But as Shapiro (1991, 13) notes, "any higher-order language and deductive system can be interpreted as a many-sorted first-order system." So this isn't a genuine alternative. The other option turns on the much contested issue of "logical constants." It's possible, as identity makes clear, to fix the interpretation of other notions as "logical" besides those traditionally fixed in first-order logic, and by providing axiomatizations (whether complete or not doesn't matter) of such notions, to give an axiomatization that could be taken to be included in the "logic" of mature mathematics. So, for example, in addition to "=," one might include "<" ("less than") plus axioms for it, or similarly for any of the other number concepts.

The normative question of what *should* be included as logical terminology, and what not, is perhaps an irresolvable one—certainly it looks that way with respect to the notion of identity. But the empirical question needn't be equally irresolvable: I suspect that the topic neutrality of the inference rules—that the recognition of logical validity occurs with sentences about all sorts of subjects and, more important, allows the recognitions of valid and invalid inferences *between* sentences about all sorts of subjects—coupled with the clear empirical origin of mathematical concepts prevents the inclusion of specific vocabulary

---

[6] The logic resulting from first-order logic with the sole addition of the new quantifier, "There are uncountably many ..."

[7] Some readers may bristle at the "therefore." My apologies: I've hardly explicitly excluded all the possibilities *here*. But I think most of them won't do as descriptions of the logic of mature mathematics—although that would take a very long survey and discussion. I've already given reasons, in part I, chapter 5, for why the paraconsistent suggestion won't do.

to that of "logic," descriptively construed as the backbone of mature mathematics.[8]

Having said this much in support of the first-order descriptive thesis, I must add two significant caveats. First, it may well be that there are variant logics that are, empirically speaking, equally satisfying with respect to mature mathematical practice; and it might even turn out that first-order logic, as we have come to formalize it, isn't—strictly speaking—among the best empirical choices (despite, let's say, being, normatively speaking, a superior choice). For example, the empirically acceptable logics may differ in their quantifier rules with respect to predicates that have the null extension.

Second, there is the important diagrammatic aspect of mathematical proof, exemplified in Euclidean geometry (as I discuss in Azzouni 2004b). Diagrammatic proof practices aren't language based, and so describing their "logic" as first-order is a category error. Nevertheless, they provide mechanically recognizable proof procedures for mathematical results. As it turns out (historically), the mathematical content of these proof procedures translated nicely into language-based systems that in turn found a place in axiomatic systems based on first-order logic. Despite this, their existence points toward the fact that, psychologically speaking, algorithmic systems that weren't language based most certainly were used by mathematicians to get results; the derivations indicated by such proofs were likely to be "constructions"—visual or kinesthetic—and not steps in language-based derivation at all. Thus, the existence of such methods of theorem proving already points toward—as in contemporary mathematics—a desertion of any particular logic as required for mathematical proof.

9.2 The Ordinary Notions of Validity, Consistency, and Consequence

In 9.1 the concern was with which logic (if any) is the logic of mature mathematics. As we've seen, this is an empirical question in just the sense that Church's thesis is: We draw an inference to the logic at work in a practice based on the form that practice—in this case mature mathematics—historically takes: whether mathematical proofs and the results of such proofs survive transplantation into a logical medium, and what alternative logical mediums (not notational variants) such a mathematical practice also survives transplantation into. But now we're turning to a description of the ordinary notion of implication: what working mathematicians (and anyone else who engages in inference) take themselves to *see* when they recognize that something follows from something else. What, conceptually, is involved (and, more important, isn't involved) in that notion? The ordinary notion isn't easy to describe as there is a problem to begin with about how to construe our usage of "consequence" and

---

[8] Actually, I have a more drastic method for addressing the problem of the logical constant. This is to deny altogether that the issue is to be resolved (or *needs* to be resolved) by sorting terms of the *vernacular* into logical and nonlogical constants. See the conclusion to part III.

"validity" as they arise in ordinary reasoning.[9] Such notions seem to involve interaction between "true" and modality like so:

(1) If S_1 is true, and S_2 is a consequence of S_1, then S_2 *must be* true as well.

(1) raises an issue faced at the very beginning of this book: Recall (1.3) the project of deflationist exegesis: All uses of "true" in the vernacular are to be in accord with its role in semantic ascent and descent—and this can be definitively determined by translation to Anaphorish. So we have here a use of "true" that calls either for elimination or for reconstrual in Anaphorish. That doesn't seem impossible to manage in this case:

(2) If S_1, and S_2 is a consequence of S_1, then necessarily: S_2.

Unfortunately, the transfiguration of (1) into (2) seems to have picked up a modal scope fallacy along the way: We don't want S_2 to be necessary *tout cout*. Another possible recasting of (1) without "true," and without the modal fallacy is this:

(3) If S_2 is a consequence of S_1, then necessarily: if S_1 then S_2.

Although we've eliminated the use of "true," and the modal fallacy, the resulting formulation is no more informative without an analysis of the notion of "necessarily."

One (purported) route to understanding ordinary consequence is via Tarski:

(4) S_2 is a (logical) consequence of S_1 iff S_2 is true in every model[10] in which S_1 is true.

Apart from the reemergence of the word "true" (although perhaps in a way that can be deflationistically tamed), (4)—and its companion definitions for validity and consistency—don't seem to capture the *ordinary* notions of consequence, validity and consistency. One reason is the involvement of models in (4), which don't—conceptually—seem similarly involved in the ordinary notion. Thus to conceptually connect the Tarskian notions to the ordinary notions, it seems we must show how model theory is implicitly involved in the ordinary concept. (But this looks like a hopeless task.)

Via the model-theoretic intrusion Tarski's approach brings, Field (1989, 31) offers two further objections to the idea that "Tarski's definitions give

---

[9] A caveat: The terms "consequence" and "validity" don't quite belong to ordinary language. The ordinary reasoner *does* talk about what follows from what (as in, e.g., "wow, that *really* doesn't follow"—something often observed after a political candidate has made a claim), what must be true given what, and what must be true (regardless). I take these as what the terms of art, "consequence" and "validity," pick out. I'll also speak of ordinary speakers being concerned with "logical" validity and "logical" consequence, although ordinary speakers don't distinguish the different ways they see that a sentence does or doesn't follow from another sentence. This, however, pretty much changes none of the issues discussed here, and so I leave detailed discussion of the matter aside.

[10] I'm going to understand models—pretty much—the way contemporary logicians do. For Tarski's original approach, and how it evolved, see Etchemendy 1990.

anything like an adequate account of the ordinary meaning of 'logically implies' or 'logically consistent'." First, such models are sets, and yet there are consistent sets of truths—say, all the truths of set theory—that are true of something (all sets) which isn't a set, and thus not a model. If such truths are true relative to some model, it must be one that "does not have the full set-theoretic reality in its domain (and in which '∈' may not even stand for the membership relation). Why on earth should anyone believe that there is such a model?"

Field's (1989, 33) second argument turns on the possibility of asserting consequences of, or the consistency of, sentences while denying the existence of the mathematical items crucial to Tarski's definitions of these ideas. That such coupled assertions don't seem inconsistent suggests Tarski's notions aren't explications of the ordinary notions but supplantations. Note that these objections both arise as corollaries to the idea that the ordinary notions of consequence, consistency, and so on, don't involve anything model-theoretic.

Yet another reason to suspect Tarski's definitions aren't explicative of the ordinary notions is that his approach allows reinterpretations of sentences in ways the ordinary notions forbid. In regard to whether "Peter jumps" (logically) follows from something, the modality of the ordinary notion presumably suggests *something* is to be varied: We're to consider the sentence relative to a range of circumstances of some sort. But Tarski's approach has too much latitude: As sentences are taken relative to different models, names and predicates may be reinterpreted *arbitrarily* provided only that the sentence remains true in each model.[11] This is an extraordinarily weak constraint which violates the identities of sentences as meaningful units: In considering various circumstances where Peter jumps, I don't want Peterless cases where Jack jumps, or Sam is a sandwich!

One last disconnect between the ordinary notion of consequence and Tarski's is, as Etchemendy shows, that the modality involved in (1) isn't captured by (4). Etchemendy (1990, 85) stresses that part of the ordinary notion is that "the truth of the premises must *guarantee* the truth of the conclusion." He continues: "However vague and poorly understood this guarantee may be, it is clearly an essential feature, if not *the* essential feature, of our ordinary concept of consequence."[12]

But is there any reason to think that either Tarski, or those who take his model-theoretic approach to these notions seriously, think of his approach as a conceptual elucidation (rather than the establishment of a Churchlike thesis linking an intuitive notion to a mathematically crisp one)? Well yes, as Etchemendy notes when considering the standard way that the importance of the completeness and consistency theorems (which show the coextensiveness between the model-theoretic notion of validity, and the proof-theoretic notion of deduction) is indicated:

---

[11] Etchemendy (1990, 23) nicely describes Tarski's approach as "in effect [involving] a characterization of '*x* is true in *L*', for a specified range of languages *L*."

[12] I agree that it's crucial to our ordinary notion; in the conclusion to part III I hazard some speculations about where that sense of a guarantee is coming from—as well as why it's "vague" and "poorly understood."

We think of these results as having an intuitive significance that goes beyond the mere coincidence of two alternative characterizations of the consequence relation. Specifically, we think of them as demonstrating the extensional adequacy of the deductive system in question. They are thought to show that the system is *sound*, that it will not allow the derivation of conclusions that are not genuine consequences of their premises, and that it is *complete*, that it allows the derivation of all the consequences of any given set of sentences in the language.[13]

On the basis of their form alone, our attitude toward the two notions thus shown to be extensionally equivalent requires an entirely even-handed attitude—but the very labeling of the consistency and completeness theorems shows that the model-theoretic notion is treated as more fundamental. This attitude (toward the consistency and completeness theorems) requires more than a Churchlike thesis—linking the ordinary (intuitive) consequence notion to Tarski's model-theoretic notion—since, via the completeness and consistency theorems, a similar Churchlike thesis is available to link the ordinary (intuitive) consequence notion to a "syntactic" notion. Only the belief that Tarski's approach reflects the ordinary notion of consequence in some fundamental (conceptual) way seems to justify the perceived asymmetry in the terminology adopted for the completeness and consistency theorems.[14]

9.3 Reasons to Resist a Syntactic Construal of the Ordinary Notion of Consequence

Before giving an analysis (in 9.6 and on) that provides an explanation for the perceived asymmetry just noted, I want to rehearse the reasons that so many have for considering a syntactic construal of the ordinary notion of consequence an obvious nonstarter. We'll see in 9.4 that these reasons don't affect

[13] Etchemendy 1990, 3–4. But, as it always is with the philosophical implications of Tarski's views, it's very hard to pin *Tarski* down to a definitive view about the relation of his notion to the ordinary one. It's worth quoting the opening paragraph of his 1983b (409) in full: "The concept of *logical consequence* is one of those whose introduction into the field of strict formal investigation was not a matter of arbitrary decision on the part of this or that investigator; in defining this concept, efforts were made to adhere to the common usage of the language of everyday life. But these efforts have been confronted with the difficulties which usually present themselves in such cases. With respect to the clarity of its content the common concept of consequence is in no way superior to other concepts of everyday language. Its extension is not sharply bounded and its usage fluctuates. Any attempt to bring into harmony all possible vague, sometimes contradictory, tendencies which are connected with the use of this concept, is certainly doomed to failure. We must reconcile ourselves from the start to the fact that every precise definition of this concept will show arbitrary features to a greater or less degree." This language doesn't nicely fit the model of establishing a Churchlike thesis, nor does it nicely fit the conceptual analysis model. It looks rather like the offer of an out-and-out regimentation.

[14] Etchemendy (1990, 4) writes: "The felt asymmetry in [the completeness and consistency theorems] stems from our assumption that the model-theoretic definition of consequence, unlike syntactic definitions, involves a more or less direct analysis of the consequence relation, and so its extensional adequacy, its 'completeness' and 'soundness', is guaranteed on an intuitive or conceptual level, not by means of additional theorems."

the Churchlike thesis I've pushed in part II, although they are rather fatal to any suggestion that the derivation-indicator view (or any "syntactic" approach, for that matter) is conceptually linked to the ordinary notion.

The first point to make is that we have (or seem to have) a strong (and confident) intuitive grip on the ordinary notion of consequence, at least in mathematics. Kreisel (1967, 153) writes: "One reasons in mathematical practice, using the notion of consequence or of logical consequence, freely and surely.... Also, it is generally agreed that at the time of Frege who formulated rules for first order logic[15]... one recognized the validity of Frege's rules."

The point is this: Frege's formalization, although differing in obvious ways from ordinary intuitions about how certain logical devices are to be understood (e.g., truth functionality, quantifiers, relevance, etc.), was recognized as otherwise largely in accord with the ordinary notion of consequence, at least as manifested in mathematics. If this hadn't been so, no one would have even seen the *relevance* of Frege's logicist project, nor the point of Russell and Whitehead's *Principia Mathematica*.

But it's precisely this sure grip on the ordinary notion which makes a syntactic construal of it seem (intuitively) so implausible. For example, Kreisel (1967, 148) notes that "both the expressive power and the limitations of first order language came as a surprise." And when he (1967, 153) considers the idea that "'ultimately' inference is nothing else but following formal rules," he writes (154) that this "is a specially peculiar idea, because 99 percent of the readers, and 90 percent of the writers ... don't have the rules in their heads at all!"

Etchemendy (1990, 2–3) similarly takes it to be "long acknowledged"

> that the purely syntactic approach does not yield a general analysis of the ordinary notion of consequence, and in principle cannot.... It is obvious, for starters, that the intuitive notion of consequence cannot be captured by any *single* deductive system. For one thing, such a system will be tied to a specific set of rules and a specific language, while the ordinary notion is not so restricted. Thus, by "consequence" we clearly do not mean derivability in this or that deductive scheme. But neither do we mean derivability in some deductive system or other, for any sentence is derivable from any other in *some* such system. So at best we might mean by "consequence" derivability in some *sound* deductive system. But the notion of soundness brings us straight back to the intuitive notion of consequence: a deductive system is sound if it allows us to prove only genuinely valid arguments, those whose conclusions follow logically from their premises.

Etchemendy (3) allows that he has made a "conceptual point" in the above: "Systems of deduction require external proofs of their extensional adequacy.... We need outside evidence that our system is 'complete,' evidence we would not require if the system straightforwardly captured, in mathematically tractable form, the ordinary concept of consequence." In both cases a stress is laid upon the resistance of the ordinary notion to syntactic construal because *conceptually* (or psychologically) it resists an analysis in syntactic terms.

---

[15] On the emergence of first-order logic, and what Frege's logic comes to in respect to that, see Moore 1980.

There are two aspects to the above uses of the contours of the ordinary concept of consequence to enable resistance to a syntactic reconstrual of it. The first we may describe as *positive* uses of it: These arise in our recognitions of when something follows from something else. A positive use is thus involved in our recognition of the validity of Frege's rules (and therefore of the relevance of Frege's program to ordinary mathematical proof). A positive use is also involved in the project of establishing Churchlike theses about syntactic approaches to mathematical proof.

On the other hand, a *negative* use is involved when we note additional aspects of some other notion, and on that basis exclude its conceptual identification with the first notion, e.g., Kreisel's remarks about the formal rules not being in our heads. The point isn't simply that we don't think of the rules when we read or write proofs—it's that whatever we seem to be doing doesn't (psychologically) involve such rules. Kreisel's remarks about the surprise upon learning of the expressive power and limitations of first-order languages are in the same category. What we're doing intuitively seems to allow us to grasp notions (*finiteness*, etc.) that first-order formalisms can't grasp; and our ordinary notion—which we're taken to be phenomenally familiar with—isn't supposed to allow the sort of surprise that we feel upon learning that the first-order formalism is as powerful as it proves to be.

There is a sense in which we can't gainsay positive uses of a concept in this sort of case; this is because, apart from one forthcoming rider, our grasp of the concept of consequence seems to literally constitute what that concept applies to. The rider is this: Part of our understanding of our grasp of consequences is that such graspings are compatible with our underestimating the involvement of tacit knowledge in what we perceive to follow from what. That is, our inferential practice (especially in mathematics, but not only there) acknowledges that our recognition that *B* follows from *A* is defeasible because we might subsequently discover that it's only with the additional assumption of *C* (say) that *B* follows from *A*. This rider doesn't allow an ultimate undercutting of the possible application of our notion of ordinary consequence—and notably, this rider doesn't allow the faulting of mathematical proofs; at worst it allows that a proof is more specialized than we may have originally recognized it to be.[16]

Negative uses of our grasp of a concept, however, can be a different matter entirely. Such uses presuppose the assumption that our grasp of the concept in question is *transparent*: that any property such a concept possesses must be one we have access to by sophisticated introspection. As an individuation condition on concepts this is, perhaps, innocuous enough: We're willing to accept, for example, that we can have *two* distinct, but extensionally equivalent, concepts (e.g., "creature with a heart" and "creature with a kidney"). There is, however, a danger of taking such individuation conditions on

---

[16] Having said this, it should be added that in practice mathematicians seem very good at recognizing that tacit assumptions are at work even when they're unsure how to make those assumptions explicit. Mathematicians, that is, easily fail to recognize *how much* is tacit—that was the theme of 7.10—but they don't fail to realize that *something tacit* is involved.

concepts too far. In particular, the view can be—as Kreisel's seems to be—that the *mechanisms* by which we recognize that our concept applies to something are transparent to us as well.

With some pairs of concepts even this isn't unreasonable. "Creature with a heart" and "creature with a kidney" have, in this sense, transparent-looking application conditions. Their coextensiveness turns on the (possibly surprising) fact that these very different application conditions pick out the same objects (in *this* world, anyway). But for other concepts this version of the transparency assumption is quite groundless. When we know a language, for example, we grasp the difference between grammatical and ungrammatical sentences. However, the mechanism by which this is done—the rules that sort sentences of a language into these two categories—is utterly opaque to us. Consider these two concepts: (i) "is grammatical," which has application conditions (for speakers of a language): sounds grammatical; and (ii) "is an R-string," where "R" describes the intricate—and introspectively unavailable—set of rules that grammatical sentences obey. These two concepts, let's say, are coextensive in the sense in which "creature with a kidney" and "creature with a heart" are; but their relationship runs deeper because how we apply "is grammatical" to those sentences it applies to is directly by way of "is an R-string," even though that fact is opaque to even sophisticated introspection. That "R-string" bears this intimate relationship to "is grammatical" can only be discovered empirically. I'll describe the concepts of "is an R-string," and "is grammatical" as "opaquely linked."

It's an opaque linkage between our intuitive notion of consequence as exemplified in mathematical proof, and mechanically recognizable derivations that chapter 7 asserted the presence of. More precisely, we successfully apply the concept of ordinary consequence to a pair of sentences—take A to imply B—because we tacitly recognize that an ordinary proof[17] of B can be constructed using A as an assumption. In turn this indicates the presence of a mechanically recognizable derivation in an appropriate algorithmic system. Unlike the case of grammatical sentences, we don't succeed in applying the ordinary notion of consequence (even subconsciously) by the mechanism of *actually* applying rules for generating mechanically recognizable derivations; that option was rejected in 8.2. We do so by our topic-relative grasp of ordinary proof; but in turn that ordinary notion of proof is (empirically) revealed to be correlated to mechanically recognizable derivations. Thus the network of concepts "ordinary consequence," "mechanically recognizable derivation," and so on, aren't linked by what, given sophisticated introspection, we take their properties to be. So in that (philosophically proprietary) sense they are different concepts. But in another sense, they are linked—theoretically linked—by a successful theory of mathematical practice. I will honor the transparency condition for the individuation of concepts: treat concepts as nevertheless distinct if they are just opaquely linked.

So far, however, what's been said has not yet directly addressed Etchemendy's initial concern: the coupled question about the significance of the

---

[17] "Proof," in the mathematical sense when the consequences recognized are mathematical ones—"argument," in a looser sense, otherwise.

completeness and consistency theorems with regard to Tarski's model-theoretic approach to the consequence relation, and the role of that approach itself. But before turning to that issue, in 9.6, I'd like to go into a little more detail about how the derivation-indicator view escapes the more immediate concerns raised in this section about syntactic approaches to consequence.

9.4 Why the Objections of 9.3 Aren't Telling against a Churchlike Identification of Deducibility (in Families of First-Order Systems) with the Ordinary Notion of Consequence

Although—as a conceptual analysis—deductive reconstruals of the ordinary notion of consequence are hopeless, the concerns that Kreisel, Etchemendy, and Field raise about the syntactic construal of the ordinary notion of consequence are easily finessed if we relax the requirement that the syntactic approach be conceptually connected (i.e., by sophisticated introspection) to the ordinary notion, and require only an opaque linking. Indeed, one of the points made in part II was that the data presented in chapter 6 about mathematical practice *aren't* compatible with a conceptual link between the notion of consequence involved in ordinary proof and the mechanically recognizable derivation of a first-order system, but only with an opaque linking.

Consider, therefore, Etchemendy's worry (quoted in 9.3) that the ordinary notion of consequence isn't tied to any single deductive system because the ordinary notion seems not to operate by means of rules at all.[18] Although this point clearly counts against a conceptual identification of the first-order notion of deducibility in an axiom system with the ordinary notion of consequence, it doesn't trouble the opaque linkage postulated by the derivation-indicator view of chapter 7, which allows a *family* of deductive systems to correspond to our intuitive grasp of, say, the standard model. In one sense, this seems to replace the ordinary notion of consequence with a family of deducibility notions; but in another sense it doesn't: Rather, there is—at best—the first-order notion of consequence which corresponds to the first-order notion of deducibility. The family of axiomatic systems—tied by coreferentiality relations among their terms—aren't tied together in any deeper sense.[19] Apart from this, not only is it unreasonable to think that specific fixed axioms in a

---

[18] In that quoted passage Etchemendy may also be raising a common way of interpreting Gödel's theorem: that PA, for example, implies certain sentences are true without such sentences being deducible from it. This issue, and how it bears on my view of ordinary consequence, is important enough to be given a section of its own (9.9).

[19] That is, it isn't that the standard model *exists*, and that somehow—from *somewhere*—it unifies the family of algorithmic systems that have terms that refer to items in it; rather, we stipulate the coreferentiality of (some of) the terms in the different systems that are compatible—given this stipulation—and that we designate as referring to items that appear in (along with others) the standard model. See Azzouni 1994, part II, §§ 6 and 7. Also see the discussion in 9.9.

specific language—apart from the first-order logical principles—are involved when we reason (for, especially in mathematics, as we saw in part II, we're always augmenting languages, although in ways that allow a conservative embedding of previously shown theorems in the richer systems), but we also switch the tools used to prove results, depending on what's easy for us and what we (happen to) know. The right picture, psychologically speaking, doesn't involve mathematicians' actually exhibiting derivations in specific algorithmic systems at all. Rather, they use all sorts of strategic devices that indicate that such and such derivations exist. This (partially) explains why the notion of ordinary consequence—when subjected to sophisticated introspection—reveals no presence of rules.

It's also worth circumventing the apparent suggestion Etchemendy makes that the "purely syntactic approach" is in danger of an illegitimate recourse to the ordinary notion should we try to conceptually link it to that notion. For, as he puts it, we're driven back to the ordinary notion when we restrict the "purely syntactic approach" to "sound" systems. This remark is slightly misleading, for it suggests that our attempts to slim down the syntactic possibilities must presuppose the ordinary notion of soundness in order to get something that has a chance of being coextensive with that notion. But it hasn't been shown that some other (sharper) characterization of a deductive system wouldn't do the job as well. Of course, we would positively (and negatively) use the ordinary notion to verify the conceptual connection between it and another notion; but that happens in any case when we try to show that two notions are conceptually coextensive.

And indeed, as we saw in chapter 6, many factors constrain which algorithmic systems constitute mature mathematics; on the other hand, contemporary mathematics is open—so I hypothesize—to any algorithmic system we want, provided only that the system is interesting. There is a sense, of course, in which we *are* driven back to the ordinary notion of consequence; but this isn't because a syntactic construal of it fails through insufficient specificity. Rather, reasoning in general, and mathematical reasoning in particular, takes place in the vernacular, and must do so according to ordinary notions of consequence. This is entirely compatible with an empirical result that our practice is nevertheless in thrall to the indication of mechanically recognizable derivations. However, because the movement from one algorithmic system to another is so unconscious and tacit, it gives rise to the impression that we're operating within the confines of unchanging subject matters (e.g., the intended model of PA) and thus we're operating with one consequence relation—which transcends any particular deductive system—rather than it being that we're switching from one such system to another. The appearance that we're operating with a single consequence relation, one that's unattached to anything deductive, is an illusion of the tacitness of the presence of the algorithmic systems, and the derivations within them, that proofs in our mathematical practice indicate.[20]

---

[20] The tacitness—and shiftiness—of (some of) the mathematical principles we're presupposing directly contributes (phenomenally) to the impression that something *about* the objects the mathematical statements concern is driving our inferences. This is something—I think—that is operating (phenomenally) in inference generally.

9.5 The Opacity to Introspection of the Rules by Which We Reason

There is a more dramatic way of describing the tacit presence of the algorithmic systems, and the derivations within them, that the proofs in our mathematical practice indicate: *The rules that are the standards for good and bad inferences are introspectively invisible to us.*

To make clear what I'm claiming, I should start with a contrast between two ways in which the principles by which we reason are introspectively invisible. One is well known, and (I should think) hardly controversial. This is that the algorithms that we employ to solve various cognitive tasks are introspectively opaque to us. Indeed, it seems clear, both from empirical studies in linguistics, and from those of our broader cognitive powers, e.g., perceptual capacities, that quite sophisticated algorithms are often employed by us "subpersonally"—that is, by substantial neurophysiological parts of us that nevertheless we haven't any conscious access to the workings of.[21] The remarks made in chapter 7 about the neurophysiological evidence bearing on the algorithms we actually execute to play games, do mathematics, and so on, are pertinent to this issue. What we do, even to mimic simple counting algorithms, proves to be (unexpectedly, and subpersonally) complicated.

In 6.5, I drew a distinction between algorithms mimicked and algorithms executed. The preceding paragraph concerns only algorithms executed. But when I suggest that the logic of mathematical proof is itself introspectively opaque, I'm no longer speaking of algorithms executed but of algorithms mimicked; and this makes the opacity claim a different, and perhaps more controversial, one. I'll try to explain why by using an analogy with games.

When we learn a game we learn a set of rules; in learning these rules, it isn't that we mentally imbibe the rules and then execute them mechanically. If that were the case, then everyone would play games (pretty much) with the same skills. Rather, we (each) bring our own individualized toolkits of algorithms to bear on the rules of the game, and in this way figure out various ways to mimic the rules of the game with varying degrees of success and failure. The point of this analogy is that the rules of the game—as long as we're playing the game—operate *normatively*. They are the touchstones used to indicate when we've made mistakes and when we haven't: when we've operated in accord with them and when not.[22]

---

[21] I avoid specific citations from linguistics, cognitive science, etc., simply because the pertinent literature is rampant with them; the point, nowadays anyway, *ought to be* an empirical truism.

[22] "Normative," as a term is tricky—and so is its purported subject matter. Here the term isn't being used in quite the same way as it was used in 9.1. We are within the context of the *descriptive* first-order thesis, as it was described there—but given that thesis, we're noting the "normative" elements first-order logic thus brings to ordinary mathematical practice as a standard of good reasoning. Talk of "normativity," perhaps ambiguously, both arises when we allude to a standard, and when we consider changing that standard.

Given this model, it might seem to follow that although—when playing a game—I can fail to have introspective access to the algorithms my mind is working through in order to enable my operating in accord with the rules of the game (to the extent that I can), it's not possible for me to not have introspective access to the actual rules of the game I'm trying to get my practice to accord with. How could those rules function normatively, as correctives to my practice, if I couldn't explicitly refer to them? Logical principles, given this analogy, aren't descriptions of how we *do* reason: They are descriptions of how we reason at our best—how we *should* reason.[23] But if logical principles are to operate normatively in this way, how can they do so without our referring to them *officially* (as it were)?

As it turns out, for logical principles to function normatively—as standards of good reasoning—they are no more required to be mentally explicit than the rules of a game operating as standards of appropriate play require *their* being mentally explicit. A small child can successfully learn to play chess and beat others at chess by merely watching, and then imitating, her elders. She needn't, as a result, be capable of stating the rules of play, or even know exactly what "rules" are. Similarly, we can separate grammatical from ungrammatical sentences, and correct our own prose (by our own lights) without being able to set down the rules we use to do this. (Linguists are still trying to figure out what those rules are!)

It's perfectly clear that mathematics—as practiced—involves logic normatively (and not merely descriptively). Mathematicians make mistakes in reasoning regularly; the important point is that they correct them (recollect 6.3). This ability to correct their own reasoning, and that of others doesn't require that the correction be done by explicitly setting out the rules of logic and testing their inferences against them. Indeed, in practice, a mathematician realizes suddenly that he or she hasn't established that B follows from A (in a proof), and now—come to think of it—maybe B *doesn't* follow from A. (Or perhaps someone points it out: "Does *that* follow?" "Well, you know, I thought it did, but now ...") None of this requires either being explicit about the rules or even being *capable* of being explicit about the rules.

I've been drawing (and distinguishing) analogies between the way that first-order logic, and its accompanying mechanically recognizable derivations are opaquely linked to our intuitive grasp of consequence, and how the set of grammatical rules by which we recognize grammatical sentences is linked to our intuitive grasp of grammatical sentence. One disanalogy is this: To say that the rules of logic, especially as they operated in mature mathematics, are introspectively invisible isn't to say that they must stay that way in all cases. Obviously we can learn them—some of us have—and learn how they're involved in ordinary mathematical proof. Furthermore, we can become totally

---

[23] The language here is dangerously slippery: Of course these principles *do* describe how we reason: They do so when we don't make any mistakes, anyway. And, this is the other marvelous thing about mathematics: Although many mistakes are made, the final result is often strikingly mistake free—at least where it matters. (Recall chap. 8, n. 24.)

aware of their operations in certain (short) proofs. I assume that nothing quite like this is possible with the grammatical rules opaquely linked to our intuitive sense of what's grammatical because in the latter case such rules are psychologically instantiated in us, and in a way that I presume is sealed off from introspective access. (We cannot introspect, as it were, our brain actually chugging through the rules and generating answers: *grammatical, not grammatical.*) But the derivation-indicator view denies that the logical algorithms are subpersonally instantiated in the way grammatical rules are. So, in learning logical rules, we aren't gaining access to how we (subpersonally) recognize valid inferences. Of course this isn't to allow that we can recognize at an introspective glance that our proof practices in mathematics indicate mechanically recognizable derivations: That can only be empirically established. But now I'm speaking of the whole practice, not of segmented bits that we can actually crank through mechanically. (I say a bit more, momentarily, about this contrast between our local grasping that a particular short proof really indicates a mechanically recognizable derivation, and our recognition that this fact applies globally to mathematical proofs.)

I've been exploring one reason why some might find controversial the suggestion that the logical principles by which we reason in mathematical proofs are introspectively opaque, despite its empirically uncontroversial cousin about the rules by which we recognize grammatical sentences. Another reason for why the claim might seem *philosophically* radical is because reasoning (in contrast to syntactic processing, for example) impresses us—intuitively—as something that *should be* transparent to consciousness.[24] This is one explanation for the perennial lure of the a priori, and the accompanying temptation to identify the a priori—suitably hedged—with the metaphysically necessary: That a proposition or sentence strikes us—intuitively—as unfalsifiable, and then in turn, as something metaphysically binding on the universe around us (so that it's seen not as, at best, an innately endowed, neurophysiologically contingent, limitation on our metaphysical imagination, but rather as a profound logical insight into the metaphysically necessary) is ascribable precisely to the opacity to introspection of the rules by which we reason. For when (psychologically speaking) we haven't access to the rules establishing a result, but only to the result itself, we *can't see* how it could have been otherwise.[25]

---

[24] In both senses of "transparent" that were discussed in 9.3.

[25] I've used the opacity of inference in Azzouni 2000a to explain why mathematical results—both pure and applied—are surprising. There I focused on the trouble we have recognizing—ahead of time—what follows from what. But that can be seen as a matter—as I presented it there (and following an old tradition)—of being overwhelmed by numerous inferential steps. But the opacity of reason—I'm suggesting here—runs deeper. One interesting thought in this respect: We (intuitively) can't distinguish between when we're reasoning correctly, and when we're strategizing, taking shortcuts which needn't be truth-preserving. This explains why we can't tell whether the powerful relevance intuitions we have are hard-wired pragmatic strategies adopted for convenience, or a manifestation of the actual rules of inference that happen not to apply in the special case of mathematics; we can't even tell whether the distinction just made is a spurious one! No resolution of this is available through introspection—through intuitions.

An objection to the opacity view of logic that one might raise at this juncture is that the subject matter, despite its difficulties, isn't all *that* alien to students. Wouldn't such principles seem inaccessibly incomprehensible if the opacity of reason, that I'm arguing for, were the case? No. The point isn't that *particular* instances of reasoning, even patterns of such, are obscure to us; that's not so. What's opaque to us is the adequacy—i.e., the global adequacy (even in the specific case of mathematical proof)—of the reasoning made available to us by a logical system. This (global) insight is much harder to grasp, and isn't introspectively transparent to *anyone*. It's here the analogy to Church's thesis, that I've been pressing in this chapter, is most pointed: It's always possible for an ordinary bit of reasoning, a bit that we ourselves have used in other circumstances, to be a counterexample to the claim that ordinary reasoning can be (directly) captured by the resources of a particular logical system.[26]

It's clear Kant, who thinks the principles by which our logical acumen operates are conclusively codifiable, bases his confidence on the (in principle) introspective accessibility of those principles.[27] His confidence is only the most sophisticated flower in a long tradition of seeing logical principles as *obvious* trivialities—this picture of the principles of our reasoning is so *very* seductive: It's the old lure of getting something for nothing.

I can't resist quickly adding one other way the opacity of reason has had an impact philosophically. Such opacity allowed some philosophers and logicians, Russell, Tarski, and Quine—for example—to essentially *dismiss* the question of what principles *really* govern ordinary reasoning. Quine (1975c, 106), notably, drives one horn of the dilemma against conventionalist views of logic via the rationale that accounts of convention that treat it as not explicit and not deliberate are *uninformative*.

Given the opacity to introspection of our logical principles, we should correspondingly distrust the significance of intuitions about what is and what's not (conceptually) included in our notions, especially with respect to the ordinary notion of consequence. We shouldn't distrust, of course, the positive use of our concept of consequence: what we take to follow from what; but even

---

As I've argued, the history of mathematics is an illuminating illustration of the opacity of reason. Practitioners were impressed very early by that "special compelling something" so manifest in mathematical proof. But such proofs, until extraordinarily late in the tradition, occur in ordinary language with practically no augmentation by technical vocabulary, a language—therefore—which totally masks the algorithms at work. No wonder that the recognition of mathematical truths—of theorems, actually—seemed for so long to be a matter of just "seeing something" about abstracta as opposed to (syntactic) insights that rules (of a certain sort) compelled a conclusion. I should also stress again how the possibility of "alternative logics" only arose once logic itself was made explicit in artificial languages.

[26] E.g., consider Boolos 1998b. Note the point: it's not that illustrations of ordinary reasoning can be found that elude first-order logic, or that ones can be found that elude higher-order logics too. The point is that such examples must be found by clever practitioners of both the vernacular and the logical systems in question; their existence (or nonexistence) isn't *transparent* to us.

[27] See the epigraphs from Kant that part III opens with. It's *this* Enlightenment view, I think, that has taken the longest to die (at least among philosophers). (Indeed, I'm working hard to kill it *right now*.)

there we must be careful about whether what we recognize as following from what is a matter of the logic, or something else entirely; and of course, even whether this distinction is genuine.[28]

9.6 Semantics 101: The Truth Table

In chapter 7, I tried to establish that reasoning, at least in the mathematical case—but this, I'd like to think, holds more generally[29]—operates so that it is correlated to systems of mechanically recognizable proofs.[30] The questions that still remain are these: First, exactly how does the opacity (to introspection) of the (syntactic) principles used to reason give rise to the powerful intuitions of *semantic* implication? Second, exactly what relation does that notion have, if any, to Tarskian-style model-theoretic approaches? And third, given views about these things, what—if anything—does the completeness theorem show?

Let's start with a really simple example of semantics: the truth table. In light of part I, we may, in fact, wonder whether the idea of a truth table with varying truth values is a notion that truth deflationists—those who take truth to be solely a device for semantic ascent and descent—have a right to. A version of this worry arises prior to the literature on deflationism; I'll follow Etchemendy's presentation of the problem:

Etchemendy (1990, 15) considers a truth table in which we vary the truth values of the three sentences "snow is white," "roses are red," and "violets are blue." It's natural to think of such a truth table as providing "truth relative to a model."[31] The question that Etchemendy poses (following Davidson 1984a) is this: What's the connection, if any, between this relational notion of truth and the ordinary (monadic) notion of truth?

Etchemendy (17) poses the problem (of connecting truth relative to a model to truth itself) this way:

> If we could not pinpoint some implicit parameter in our ordinary notion of truth, some parameter whose potential effect on the "absolute" truth values of our sentences is mimicked by the effect of changes from row to row in the theory of

---

[28] One philosopher who clearly takes this distinction to be one we project onto the phenomena for purposes of sheer regimentation is Quine (see Quine 1960, e.g., 11).

[29] Recall, however, chap. 6, n. 32.

[30] As I keep saying, this is an empirical matter. It's important, though, to stress that we became convinced of the codifiability of mathematical reasoning within deductive systems with such and such properties, not on the basis of semantic arguments, but because of a translation project from the vernacular into that logical system (e.g., *Principia Mathematica*). And our recognition of logical validities, dating from much earlier, was via recognizable classes of argument schemata, e.g., in Aristotle. The grip of semantics—formal and otherwise—is *really recent*, given the history of mathematics and logic (in particular); and even though the way recognizable classes of argument schemata were characterized: argument-forms—the instances of which remain true however "nonlogical" terms in the sentences are replaced by (grammatically equivalent) terms—is the clue Bolzano and Tarski use to develop their semantic approach.

[31] In contrast to Etchemendy, I claim: The models *are* the rows: We can call it: "truth in a row," or "truth relative to a row." See what follows.

relative truth, then Davidson would be completely justified in claiming that the defined 'x is true in y' is irreducibly relational. And consequently, he would be justified in claiming that, for this reason, our theories of relative truth cannot be thought to illuminate the notion of truth as we ordinarily understand it.

But the notions *are* connected, although there is a choice between "two obvious alternatives":

> There are only two parameters to which the sentence 'Snow is white' owes its truth: broadly speaking, the *language* and the *world*. . . . [H]ad the language been somewhat different, this sentence would have been false *in spite of* the whiteness of snow. . . . On the other hand, had the *world* been different, this sentence might have been false in spite of its *meaning*.

We do one of two things by moving among the rows of the truth table: We imagine variants in the world (e.g., metaphysically possible worlds where roses are blue, or violets brown) or we imagine variants in the language ("roses" instead refers to violets, "is red" holds of, instead, only and all, books). This optional fork is extremely natural—there really seem (at first glance) two and only two possibilities here;[32] but nevertheless this really is an *evil fork* (that Etchemendy is sweetly offering us for purposes of impalement), and it's one the truth deflationist should (and can) reject, even if that deflationist is *not* a metaphysical truth deflationist. For either we must construe models (here the alternative rows of the truth table) as metaphysical models of how the world can (and can't) change—and face corresponding concerns with whether the inscrutable metaphysics of the universe matches up with the possibilities and necessities of logic—or, on the other hand (as I've already noted in 9.2), we must construe changing the truth table rows as (strangely) changing the semantic identity of the sentences.

Notice the problem facing the truth deflationist: He wants to remain neutral about how sentences are true and false—in particular he wants to remain neutral about whether the truth or falsity of a sentence turns on how the world might be different. But then, so it seems, in rejecting Etchemendy's first option, he must adopt the second; or he must fail to capture the semantic notion of a truth table altogether.

9.7 How the Truth Deflationist Acquires the Notion of a Truth Table

Luckily, there is another route for the truth deflationist. To see it, let's return to the sort of example Tarski (and Bolzano before him) drew the inspiration for their approach from. Consider a standard syllogism:

---

[32] Etchemendy (1990) develops this fork into two alternative approaches to model theory: representational semantics, in which the semantics turns on variations on how the world can be, and interpretational semantics, in which the semantics turns on variations on reinterpretations of the nonlogical parts of sentences, and he shows that Tarski's approach can't be justified on either view—that it apparently draws its motivation sometimes from one and sometimes from the other.

All **A** are **B**
All **B** are **C**
Therefore   All **A** are **C**.

As the deliberate use of schematic letters indicates, this argument form doesn't depend on what's substituted for "**A**," "**B**," and "**C**." The insight, therefore, is this: There are *two* phenomena at work. One is *inference*; if the premises are true, the conclusion must be also. The second is that the mechanism of inference—whatever it turns out to be—doesn't depend on the specifics of "**A**," "**B**," and "**C**." However—without argument—there is no reason to think these two facts need be intimately *connected*.[33]

To say that the conclusion *follows from* the premises *because* substitutions for nonlogical vocabulary don't affect the validity of the inference looks like a *non sequitur*. Why should substitutions—why should inferences involving *other sentences* entirely—have any relevance to the validity of the original argument? What the substitution fact *does* offer is a clue and a hope. The hope is that one can say something *general* about inference; one isn't stuck with lots and lots of specific valid inferences with nothing in common. *Codification* of our logical principles looks *tractable*. The clue is that whatever is going on with ordinary validity turns only on specific terms in the language and not on all of them.[34]

It may seem this project of the codification of our logical principles requires a *principled* distinction between logical and nonlogical vocabulary in the vernacular; but making *that* distinction out looks hopeless. We need to (again) carefully distinguish the normative concern from the empirical one, as we did in 9.1. Just as whether and what syntactic rules of inference are operating in ordinary reasoning is an empirical matter, so too what sorts of vocabulary are the *primary* logical linchpins in that vocabulary is similarly empirical. For our current purposes, I take it as empirically established that for significant chunks of the reasoning in the vernacular, including what's utilized in mathematical reasoning, the primary logical linchpins can be empirically established as (transliterations of) the logical idioms of first-order classical logic. This *isn't* right;[35] but what *is* right isn't different in pertinent respects, and so for illustrative ease, I show in terms of the simpler view how the

---

[33] Indeed, Etchemendy (1990) shows that Tarski's interpretational approach to consequence comes apart from the ordinary notion: Even if there proves to be an extensional connection between these notions, so that we can define, for example, a sentence as logically valid iff its truth is preserved under all (suitably hedged) substitutions for its "nonlogical" terms, this doesn't reveal what it is that guarantees the sentence *must be* true.

[34] Tarski (1983b, 414), borrowing very old language, says, "We are concerned here with the concept of logical, i.e. formal consequence, and thus with a relation which is to be uniquely determined by the form of the sentences between which it holds [;] this relation cannot be influenced in any way by empirical knowledge, and in particular by knowledge of the objects to which the sentence[s refer]." This sort of (motivational) talk is extremely misleading.

[35] Why not? We've seen this: The devices in ordinary language—the quantifier-expressions in natural languages, for example—are far more complicated than the devices of the first-order predicate calculus. I suspect that no vocabulary in ordinary language functions solely as the logical linchpins for our inferences, especially since those inferences aren't themselves mechanically

Tarskian approach to validity and implication proves relevant to the ordinary family of homophonically named notions.

Let's return to the ordinary notion of consequence, as expressed in (1), here repeated

(1) If S_1 is true, and S_2 is a consequence of S_1, then S_2 *must be* true as well.

The striking fact about the modal notion here—and this holds of modal notions generally when they arise in ordinary language—is that although it does seem given to introspection or intuition that the necessity in question is relative to the varying of something, what precisely is supposed to vary *isn't* clearly given to introspection or intuition. Many philosophers are prone to take a metaphysical leap of faith about this—just as we saw Etchemendy do when he offered his readers the fork a while back. But a metaphysical interpretation isn't part of the intuition of the modality—and that's why metaphysically deflated views of modality don't seem intuitively implausible. Indeed, carefully looking at (1), all that seems to be implicitly taken as varying is the truth (values) of sentences.

Thus, although the varying of *something* is crucial to the modality in (1), it needn't either be the world or the content of the nonlogical terms in S_1 and S_2 that so vary; another (overlooked) option—but one quite clearly natural given the ordinary-language formulation (1)—is this instead: S_2 is an ordinary consequence of S_1 if, regardless of how the truth values of all *sentences* are varied (subject to certain caveats), S_2 is never false when S_1 is true.

Two (related) questions immediately arise: First, is this approach to ordinary consequence one the truth deflationist has access to? And, second, is this approach open to a metaphysically deflated construal, so that the MTDist of part I can also find it acceptable?

The answer to both questions is "yes." To begin with, recall from 1.1 that the truth deflationist has access to the notion of the extension of the truth predicate.[36] Thus, the variants in question are to turn only on changes in what's in the extension of the truth predicate (they aren't to turn—at least not directly—on *how* what's in that extension got there; e.g., no assumptions need be made that the truth predicate could contain something different only if the world itself could differ in certain respects from how it actually is).

For example, it may be that "$2 + 2 = 4$" is metaphysically necessary, and *can't* be false. No matter: *Metaphysical necessity is not logical necessity*, and so there are possible extensions of the truth predicate without "$2 + 2 = 4$." So too, it may be that there is (metaphysical) necessitation between snow's being

---

recognizable derivations but only indicate such. This topic will be discussed further in the conclusion to part III. For now, I'll note that only a view which treats our ordinary language inferences as either exemplifying actual derivations or as being strict abbreviations for such requires a sorting *in ordinary language itself* of the logical from the nonlogical vocabulary.

[36] I've been speaking of truth deflationists generally here rather than of biconditional truth deflationists because even a truth deflationist of my stamp—one who assimilates the truth idiom to anaphorically unrestricted quantifiers—still has access to a truth predicate; recall 3.2.

white, and water's being transparent. *Whatever:* If it isn't a matter of *logical* necessity (and it's not), there *can be* extensions of the truth predicate containing one of those sentences but not the other. More radically, it may be that a sentence is true, but not because of how the world is (since that sentence hasn't got terms that refer to anything that exists); again, no matter. The extension of the truth predicate can vary with respect to the presence of such a sentence without worrying about how the world does or doesn't change as a result.

Because the ordinary truth deflationist has access to "true" as a predicate, as something that has sentences in its extension, he similarly has access to the notion of variation in the contents of that extension. Indeed, translation into Anaphorish, as we can see from (3) (in 9.2), is possible: The anaphorically unrestricted pronouns just need to fall within a necessity operator; but that's glossed—nearly enough—as variations in certain sentences given *other* sentences (or not). And, as we've just seen, the truth deflationist's metaphysical neutrality (recall 1.1) is untouched by the possibility of so varying the contents of the extension of "true" precisely because of the metaphysically deflated interpretation of that varying (that varying sentences in the extension of the truth predicate doesn't require an analysis of how the world can't or can be—metaphysically speaking—different).[37]

What are the constraints, if any, on the varying of the extension of the truth predicate? Well, this is, as already indicated, a matter to be empirically established: For our (illustrative) purposes, there are no logical connections between any of the names and predicates of the language; all logical connections between sentences arise only from designated logical vocabulary: connectives and quantifiers. This means that sentences without logical particles within them are logically independent of one another (and this logical fact, again, is entirely independent of metaphysical facts that can tie sentences together by virtue of what they're *about*).

I'm glossing, I hope it's clear, the *ordinary* notion of consequence (a conclusion—in a valid argument—holds come what may, provided only that the premises are true) in terms of a "come what may" of *sentences differing in possible truth values.* But, as we've seen, metaphysical overenrichments of the idea are all too easy to slide into—and not just on the part of philosophers. Why? First, the opacity of the rules of inference (specifically, an inability to introspect the centrality to our reasoning of certain linchpin terms—connectives, quantifiers, and such)[38] coupled with a (mistaken) correspondence truthmaker picture of truths (i.e., truths are and must be *made true* by what they're about) leads to a construal of "come what may" in terms of *items in the world and how they can differ from what they actually are.* But this (in

---

[37] One caveat: *Snapping* the connection between metaphysical possibility and necessity *and* logical possibility and necessity raises the specter of a *mismatch* between the two: Language, or the logic embedded in it (anyway), may be metaphysically misleading. That's right! Metaphysics is really *hard* because we don't have the methodologically cheap option of reading metaphysical pronouncements directly off of our logic.

[38] This is why the discovery of argument-schemata by Aristotle was a *discovery*.

turn) leads to a conflation of logical possibility (and necessity) with meta-physical possibility (and necessity). And that, in turns, leads to *outrageous* amounts of epistemic woe and sorrow (i.e., to hopelessly irresolvable questions about how *we*—fallible animals with *really poor* imaginations—manage to fix *our* logical principles by the invisible *light* of implacably grand metaphysical facts).

9.8 Connecting the Ordinary Notion of Consequence to the Tarskian Notion

Given that I'm right about the ordinary notion of consequence, how does the Tarskian approach prove illuminating of that notion, and others in its family, e.g., consistency, and so on? Well, it helps to recall the Henkin proof of the completeness of the first-order predicate calculus. What's shown, typically, is that given a consistent set of sentences, a model can be constructed for it. That model is constructed by first embedding the consistent set of sentences in a complete maximally consistent set of sentences,[39] in a language-extension of the original language of the sentences, one that contains (at most) a countable number of additional constants.[40] This resulting maximally consistent set of sentences, in turn, is used to construct the model *literally* out of the new (and old) constants themselves (which serve as the domain of the model).[41]

This fact—and indeed, one of the corollaries of the completeness theorem for first-order logic (Löweinheim-Skolem, nearly enough), that such flimsy stuff (from the set-theoretic point of view) as countable collections of constants, n-place predicates defined to hold over n-tuples of the constants, and constants stipulated to refer to constants, *suffices as a model* for any consistent set of sentences—once upon a time was the bane of (certain) philosophers.[42] But for ordinary truth deflationists seeking to characterize the significance of Tarski's approach for ordinary consequence and validity, it's nothing of the sort because models are *not* meant to represent alternative ways the world can be (wherein *somewhere* among these the actual world appears, or at least an isomorphic mirror image of it does); no, all that's being represented are the different ways logical constraints between sentences allow (every) *sentence* to be true and false independently—or not—of one another. Thus model theory is sheerly *picturesque*: It encapsulates the hypothesis that, from a logical point of view, all that's relevant about the world (to the truth and falsity of various

---

[39] That is, sets of sentences of the language wherein every sentence or its negation appears, and wherein not both a sentence and its negation appear.

[40] The additional constants are needed as witnesses: that statements of the form: $(\exists x)Px \Rightarrow Pa$ can be added to the consistent set of sentences, for every formula: $(\exists x)Px$.

[41] Such proofs are available in nearly every logic textbook. A nice presentation may be found in Hodges 1983, esp. 60–65.

[42] Putnam (1983, 1) begins with a description of Skolem's result as "an antinomy, or something close to it, in *philosophy of language*." The Melodrama of the Antinomy, however, turns crucially on taking Tarski's model theory (n. 32) representationally. Putnam's antinomy *evaporates* under my forthcoming construal of the role of model theory in relation to ordinary consequence.

sentences), *if anything is,* is the interaction between objects and relations: Nothing else matters, and so sentences (without logical particles) are made true by "objects" having the "properties" corresponding to predicates correctly appended to sequences of terms.

I can call (get away with calling) model theory "picturesque" because, on this view, it isn't required that there actually *be* any such objects or properties. I mean this in two senses. First, and this is the point of the "if anything is" italicized in the last paragraph, the language this semantics is for can be one where the sentences in question are about things *that exist in no sense at all.* Nevertheless, from a logical point of view, the same thing is going on whether objects and properties *really* exist, or neither do. The correct appending of predicates to sequences of constants, and quantifications involving predicates and variables (e.g., instantiation claims)—both of these don't require anything real: The ontic question is irrelevant to the logical one of how the truth values of these sentences connect to the truth values of other sentences.[43] But second, *model theory itself,* and the set theory it's part of, needn't be ontically committing in any sense at all—*nominalists* can happily help themselves to set theory—as it stands—and thus to model theory as well. Thus the objects and properties impounded for purposes of Tarskian model theory aren't ones we need be ontically committed to.[44]

With these points in mind, let's return to the issues that have arisen in the course of this chapter, ones raised (in large part) by Etchemendy, Field, and Kreisel, to see how they can be handled. First, there is the threat (recall 9.6) of a potential disconnect between the monadic notion "true" and the relational "true-in-a-model." Also recall (9.2) Field's concern, which is related to this first point, that the actual set-theoretic universe, say, isn't itself a set, and thus can't function (directly) as a model. We now have answers in place: The proxy of the sort that Field (1989, 32) contemplates, and fears doesn't exist, isn't *needed.* All that's required is that the *sentences* true in (some) model are the sentences true of set theory; and that this happens is shown by the completeness theorem (coupled, of course, with our empirical assumptions about the language of set theory). It follows—as I've mentioned—that the mathematical richness of standard model theory is (pretty much) of no interest to the ordinary notion of consequence. It's thus of no interest philosophically or, for that matter, logically: It doesn't contribute to our understanding of validity, consistency, and so on, for example—unless one slips into metaphysical construals of the role of model theory that I've (here) inveighed against.[45]

---

[43] Objectual quantification *is* simply the connection, via semantic conditions, of quantifiers in the object language to those in the metalanguage—not a connection, I must stress, that need be to anything that's real. (At least not on the mere basis of a connection via semantic conditions between quantifiers in two languages.) See Azzouni 2004a, 53–55.

[44] Here too Azzouni 2004a is relevant, especially chapters 3 and 4.

[45] Saying this, of course, doesn't fault in any way its genuine *mathematical* interest. But that's extremely common: We often think a mathematically rich application to something empirical has (as a direct result of its sheer *mathematical* richness) philosophical ramifications when we overlook what the mathematics is really doing.

And this, in turn, explains why the reinterpretation issue that I raised in 9.2—that the Tarskian approach allows sentences to strangely vary in their semantic identities from model to model—is also of no concern. From a logical point of view, the actual referential relations so crucial to the semantic identities of sentences (on some construals of "sentence") are of no significance in trying to capture the notions of validity and consequence understood as variations in what's true and false. All that matters is the connections between sentences, and the *attributions* of "relations" to "objects."[46] But such attributions, and sentence-to-sentence connections, are insensitive to the *particulars* of the actual objects referred to. What Tarskian model theory *is* sensitive to (and all it *needs* to be sensitive to) is that the relations defined in a model hold of objects in that model so that sentences are appropriately assigned truth values. This abstracts away from the other (possibly semantic) aspects of sentences, but not in a way that endangers how such sentences—understood fully, semantically speaking—can vary in truth values.[47]

This last point is related to something else that I've already addressed, but perhaps not explicitly enough. It's often pointed out (e.g., by Etchemendy 1990, chap. 8) that the (indispensable) set-theoretical assumptions of Tarski's approach seem to make logical truths of statements, e.g., about the size of the set-theoretic universe, that should *not* be matters of logic.[48] But this worry is *doubly resolvable*. First, there is the point, already made, that the adoption of set theory is (in any case) ontically innocuous: The nominalist can use the resources of set theory, with all its standard "existential assumptions" just as any flamboyant Platonist does; that nominalist just won't interpret the existential quantifications appearing in set theory, and used in the application of that set theory to semantics, as *ontically committing*. But second, even if we table this position (of mine), the resulting models that the Henkin completeness proof supplies show in any case that very little is actually required of model theory to underwrite the notion of validity that the ordinary deflationist wants: Once the need for set-theoretic proxies of the universe is recognized as spurious, countable first-order languages need only, at best, countable universes to facilitate the Tarskian approach to consequence and related notions.[49]

This brings us, finally, to a deep methodological issue about the application of mathematics.[50] Etchemendy (1990, chap. 8) worries that the notion of ordinary consequence is dangerously in thrall to substantive facts about the

[46] I've generally, for reasons of style, avoided enclosing terms in quote marks as a device for "bracketing ontic presuppositions," but here they seem especially useful.

[47] Again, all this is subject to the empirical assumption that only the logical linchpins codified in first-order logic are the ones relevant to the ordinary notions of consequence, validity, etc.

[48] The polemical assumption at work in this argument is one I otherwise agree with: Ontic commitment isn't to be dictated by the principles of one's logic.

[49] An irony of this last point is that the Löweinheim-Skolem theorem that those enamored of categoricity inveigh against proves to be an advantage: It explains why the Tarskian approach—despite its top-heavy mathematics—*really is* an elucidation of the ordinary notion of consequence.

[50] Oddly, it's on *this issue* that I most strongly part ways with Etchemendy, Field, and Kreisel. And yet, as we'll see, the difference in opinion turns on an almost minor switch in perspective.

set-theoretic hierarchy: Imagine that the set-theoretic universe were finite, then Tarski's approach would declare wrongly on the logically true, the consistent, and so on. But, really, all this amounts to is that in *that* case, set theory would have proved to be the wrong branch of mathematics to apply empirically to the phenomenon of human reasoning, and we would have had to dig around in our mathematical toolkit for *something else.*

Compare it to this case: Suppose that, in fact (in Platonic heaven, or wherever such things romp), 3-space—the abstract object—*wasn't flat.* Then the application of Euclidean geometry to Newtonian physics would fail.[51] Does this mean that, in the Newtonian context, ordinary space (still) waits upon a conceptual analysis of it that's independent of the mathematics used to characterize it? It's hard to see why—except in the sheerly empirical sense that we might discover empirical phenomena (as in fact, we have!) that show a different branch of mathematics suits this empirical application better.

9.9 A Remaining Worry

One last point should be revisited.[52] The ordinary notion of consequence seems to operate by means of the notion of truth, and, in accepting this, I've labored to show that ordinary deflationist truth can be impounded for this use. However, just that use of "true" may (to some) seem to force us away, on impeccable technical grounds, from a syntactic construal of the ordinary notion. The concern, of course, is Gödel's theorem.[53] Let $\mathcal{L}$ be a language whose syntax is effective, and suppose $\mathcal{L}$ can express a certain amount of arithmetic. The variables of $\mathcal{L}$ range over the natural numbers, and we introduce a Gödel numbering: $\lceil \phi \rceil$ which assigns every natural number a sentence. A is an effective and sound theory in $\mathcal{L}$ and contains the rudimentary axioms of arithmetic. $\mathbf{PRF}(x, y)$ is the proof predicate for A definable in $\mathcal{L}$, and let $G \equiv \neg \exists y \mathbf{PRF}(\lceil G \rceil, y)$ be the fixed point for $\neg \exists y \mathbf{PRF}(x, y)$.

Now, roughly, Gödel's first incompleteness theorem is that G is true but not provable in A (provided that A is consistent). It might seem, therefore, that the ordinary notion of consequence (as formulated by (4)) licenses the idea that G is a consequence of A because: If A is true, G must be true. Indeed, as Shapiro (1998, 498) points out, to show this in an A^* augmented over A with a truth predicate, we need

(*) $\forall x(\exists y \mathbf{PRF}(x, y) \Rightarrow T(x))$,

which says that all of the theorems (of A) are true.

But something slippery just happened. It isn't that if A is true, then G must be true. In fact there are (first-order) models in which A holds and G doesn't. What's the case is that there is an A^*, which if appropriately supplemented with

---

[51] Not exactly; this would depend just on *how* it failed to be flat; but leave this aside: The example is still illustrative.
[52] Recall note 18.
[53] What follows this sentence is cribbed (and modified) from Azzouni 1999.

principles (like ($*$)), has G as a theorem. But isn't that enough? ($*$) *says that all of the theorems of A are true;* and so it's revealed in A^* that: If all the theorems of A are true, then G is (must be) true too.[54]

Let's be careful. Consider a model M of A in which G and ($*$) don't hold. In such models A is nevertheless sound; and according to that model, all the theorems of A are true as well. What, then, is wrong with ($*$)? Well, $\mathbf{PRF}(x, y)$ isn't the proof predicate in M (relative to a Gödel numbering)—or more accurately, it isn't *just* the proof predicate: ($*$) is making a claim about *all* the numbers in M, and what it's claiming about (some of) them is false.

Thus, by thinking of $\mathbf{PRF}(x, y)$ (solely) as a proof predicate, we have saddled A with a specific subject matter that, strictly speaking, it hasn't access to. When we do that, we implicitly grant A more resources than it genuinely has, and as a result, we narrow the models we take it to hold in. No wonder, therefore, it seems that if A is true then G must be true as well.

The point can be seen a little more directly if we note, as Shapiro (1998, 499) does, that "[i]f A' invokes something like Tarskian satisfaction, then it has a natural extension that [contains G as a theorem]": namely, if we introduce a truth predicate, and allow it within the induction scheme. We should do this in the case where the subject matter is arithmetic, Shapiro thinks, because, as Shaughan Lavine (1994, 231 n. 24) writes: "Part of what it is to define a property of natural numbers is to be willing to extend mathematical induction to it. To fail to do so is to violate our rules for extending and further specifying our arithmetical usage."

But here is a place where the demands of truth have come apart from our intentions toward characterizing, say, *the standard model* of PA. To characterize the ordinary consequences of PA we invoke truth; but we don't need to place the truth predicate so invoked within the confines of the induction schema. Indeed, if we do so, we are *excluding* models of arithmetic which PA can otherwise hold of, and doing that falsifies our claims about what must be true if PA is true.

---

[54] Shapiro (1998, 499) sees the reasoning in A^* as replicating the following informal reasoning: "Once our subject has taken on the truth predicate, and he notices that all the axioms of A are true and that the rules of inference preserve truth, he concludes that every theorem of A is true. . . . So our subject concludes . . . that G is true."

CONCLUSION TO PART III

I want to give an overview of what's been achieved in chapter 9 and, along the way, address a couple of worries that may still be haunting some readers. I'll do this by linking up, in a slightly different way, the chapter's themes.

Etchemendy's problem is this: We've got an ordinary notion of consequence that we confidently grasp (at least in the case of mathematics). Somehow Tarski's model-theoretic approach seems to be taken by the cognoscenti to be an elucidation of that notion. But given the technical details of Tarski's approach, how can this be?

My solution is to recognize the (surprisingly) central role of the Löweinheim-Skolem property of first-order systems in facilitating a conceptual connection between Tarski's approach and the ordinary notion of consequence. The modality of the ordinary notion seems to be concerned with varying truth values of *sentences* (outright) or, more accurately, with varying the presence of sentences in the extension of the truth predicate rather than with varying properties of objects—and those objects—in the world or with varying nonlogical vocabulary in sentences. But precisely because Henkin-style constructions of models suffice for the completeness proofs for first-order languages, the model-theoretic richness of the Tarskian approach turns out to be irrelevant to how that approach elucidates the ordinary notion of consequence (glossed in terms of the varying of the truth values of sentences). This is why the Tarskian approach strikes one as an elucidation of the ordinary notion of consequence: It actually is—if (i) one uses the Henkin proof for completeness to recognize that most of the models available are idle wheels, and if (ii) a case can be made that the inferential connections between sentences are first-order representable.

Some concluding caveats. First, my view of the relation of the ordinary notion of consequence to first-order formalisms really isn't that inference is "just following first-order formal rules." As I argued in part II, how we actually recognize a valid inference involves an intricate story about how our proof practices accord with the indication of mechanically recognizable derivations. This is a fact that mathematicians needn't be aware of—although they are always aware of the compelling quality of mathematical proof.

This emancipates mathematical proof—and reasoning in general, I suspect—from the ordinary language it seems to be couched in. The terms of ordinary language often contain additional—from the first-order point of view—semantic (and pragmatic) excrescences. For example, the terms for counting numbers in all natural languages contain peculiar semantic connections to each other that aren't mathematically systematic.[1] Similarly, terms like "and," "or," "if—then," and the various quantifier terms have all sorts of semantic and pragmatic complexities involving tense and other factors—and although some of these arise only in specific circumstances, they do so in ways that are extremely hard to systematically characterize. The mathematician, operating in ordinary language, instinctively strips these terms of this stuff and operates with them "as if" they are missing those semantic additives. Tractable implications—i.e., implications that correspond to mechanically recognizable derivations—between sentences can emerge into explicit consciousness when this is done.[2]

I suggest that this (implicit) "stripping" practice is a more primitive version of the explicit introduction of terminology into the vernacular—terminology that can be stipulated in its semantics—or, related to this, the introduction of out-and-out regimentations: artificial languages with syntax that reflects only the inferential relations between sentences that we want.

And if right, this enables the circumvention of a problem that has bedeviled analyses of the logical form of ordinary-language sentences for some time: the problem of the logical constant. The problem is that there doesn't seem to be a principled way of determining which vocabulary items should be taken to be part of the logical vocabulary of ordinary language, and which items should be assimilated to the nonlogical vocabulary. The issue has bite because what logic—first-order or otherwise—is actually exemplified by a

---

[1] Just count in English, and notice how you say, eventually, "ten," "eleven," "twelve," "thirteen, ..."

[2] Actually, more dramatically, certain implications become *possible*: The excess semantic structure of terms may disallow those implications otherwise. So the mathematician is, in this sense, deserting natural language without officially doing so. The automatic stripping of tense from mathematical language is an obvious example of the sort of phenomena I'm talking about.

One nice piece of evidence for this view is that people—even in antiquity—got onto the idea of truth-functional construals of certain notions, like "and," and "if—then" (see Kneale and Kneale 1962, e.g., 129–30). It's hard to see how they could have done so without—essentially—stripping away from ordinary-language terms some of their semantic or pragmatic roles.

language turns directly on what logical vocabulary it has. In particular, the first-order descriptive thesis (9.1) seems to come to grief precisely because the slender list of logical constants necessary and sufficient for first-order logic doesn't seem especially indicated in ordinary language: Many more items look like logical constants than the first-order descriptive thesis allows. Worse, the items in the vernacular that are closest to first-order idioms ("there is," "and," "if—then," and so on) clearly have implicational properties that deviate—sometimes drastically—from their first-order cousins.

On my view, none of this is a problem because the first-order descriptive thesis isn't beholden to there being actual items in natural languages which exemplify the first-order properties of the logical constants in first-order languages, any more than a thesis that our ability to recognize the properties of counting numbers turns on, say, grasping the axioms of PA, requires there being in ordinary language (number) terms which (semantically) obey those very axioms. In both cases, we escape whatever obstacles the semantics and the pragmatics of the terms of natural languages place in the way of our understanding of what we take to be true of the counting numbers (or what we take to follow from what) either (i) implicitly, by rarifying the usage of the terms of ordinary language and coupling that with an (implicit) agreement of how sentences in specific contexts (e.g., in mathematics) are to be understood (regardless of whether such an understanding fits with the semantics and pragmatics of ordinary language); or (ii) explicitly, by using the regimentation in an artificial language as a touchstone of what we should take to follow from what.

This provides a sketch of a solution to a problem some may have detected early on in the derivation-indicator view of mathematical practice. This is the issue of *what medium* mathematical proofs were supposedly taking place in. If in ordinary languages, then there is the worry that the semantics of those languages are too much at variance with the supposed language of the derivations indicated for mathematics to successfully make the leap. One response, of course, is to invoke the large metaderivational element to ordinary proof: to say that the practice—to a large extent—is talking about such derivations (and therefore, can do so in a language quite different from the language of the derivations themselves). But one can worry that there is still an issue about how the mathematician is supposed to get onto (even implicitly) the shape that such derivations have. Does he (implicitly) perceive the contours of derivations as they would appear in regimentations—does he do this even centuries before the idea of such things emerges into consciousness? The solution that the (tacit) stripping of (certain) semantic properties of some of the terms of ordinary language allows is that the medium in which such derivations can (in principle) appear (at least tacitly) is that of ordinary language—ordinary language so (implicitly) modified, anyway. This is a historically "healthy" suggestion—for it makes Frege's invention of artificial languages not something that emerged (historically) straight out of nowhere. Rather, it had roots in mathematical practice as it had been going on for some time.

This solution to the problem of the logical constant in the vernacular thus proves to be the companion view to the derivation-indicator theory of mathematical proof. It also provides additional support for the first-order descriptive thesis argued for in 9.1 because—as noted—it eliminates the need for a systematic division between logical and nonlogical constants in the vernacular. My assuming a division of such was available (during the course of chapter 9) is therefore an expository simplification that (ultimately) isn't presupposed by the arguments I gave there.

One last point. Recall (1), here repeated:

1) If S_1 is true, and S_2 is a consequence of S_1, then S_2 must be true as well.

I've argued that it's reasonable to gloss this ordinary notion of consequence in terms of variations in possible extensions of the truth predicate, with respect to the sentences S_1 and S_2. (S_2 is a consequence of S_1 if, (i) regardless of how the truth values of all other sentences are arranged, (ii) if S_1 is contained in the extension of the truth predicate, then so is S_2.) The extent of the variation possible (that is, *which* sentences can vary independently of each other, and which not) depends on what the logic is that we treat the sentences as beholden to. Were it introspectively accessible to us that we're treating our sentences as beholden to (say) their unabbreviated forms in a first-order formalism—in other words, contrary to reality, were all of our mathematical proofs quite short and sweet, and were the deviations from ordinary language easily seen and quite explicit—it would be easy to conceptually connect (1) with a syntactic construal of deduction. The tangle of issues we've been driven to consider in this book arise only because successful mathematics required a much more circuitous connection between ordinary cases of reasoning and mechanically recognizable derivations.

I should add that even in this idealized case, the semantic notion couldn't be implemented: It can't be seriously entertained that anyone (ever) considers (all the) variations in what can appear in the extension of the truth predicate, and in this way sees that *A* implies *B*. Only a local syntactic notion is one we can use. Thus in this idealized circumstance, I hypothesize that we would just have the syntactic construal of deduction: the notion that *A* implies *B*, if given *A* is true then (come what may) *B* is also, wouldn't even have occurred to us. It does so (in our unidealized circumstances) only because the syntactic motors of inference are introspectively invisible. This means, in addition, that in a case—contrary to fact—where we had intuitive access to the rules we really are using to infer, we wouldn't have the sense of a guarantee either. Let me take a couple of paragraphs to make this case and to add a couple of caveats.

First, when I claim that only a local syntactic notion is one we can use, I'm not claiming that when we go through a traditional proof, we're actually engaging in a formal derivation that corresponds to it according to the derivation-indicator view. That derivation is arrived at only when everything in a proof has been made explicit. Rather, in practice, several things are going on. Some steps are explicit (e.g., modus ponens, the shift from a general statement to a specific case, and so on), but many others encapsulate numbers of tacit assumptions bundled together in inference steps that mathematicians take at once. Indeed, this kind of encapsulation of assumptions is part of what makes mathematical proof appear to be semantically driven—to be based on aspects of the objects that the proof is about.

Second caveat: It's important to realize that we usually recognize that *A* follows from *B* not by seeing that, regardless of the truth values of all other sentences, if *A* is true, then *B* will be true, nor by seeing that *B* follows from *A* by means of a syntactic rule that we recognize ourselves to hold as valid. We do so by the guidance of a feeling that we can't *imagine B* being false while *A* is true. This is why we can get it wrong—mistake a failure of imagination for something else. In some cases, and modus ponens may well be one of them, we can simultaneously be aware of our being unable to imagine otherwise, and that a mechanically recognizable rule is being applied—although, and this is significant, our sense of the source of the validity of the inference is our being unable to imagine how it could be otherwise, and not its mechanical recognizability.

I should, however, say something more about the strange way that this intuitive sense of a guarantee arises in us. I'll start by stressing that the benign fixation of mathematical proof that was discussed in chapter 6 should be distinguished from this intuitive sense. Some might think that the intuitive sense of a guarantee that we feel when we carry out the steps of a proof is the source of the benign fixation of mathematical proof. This can't be if only because so much of mathematical reasoning is based on assumptions that we have discovered to be ones we can reject. In whatsoever sense the original axioms of Euclidean geometry may have been seen as guaranteed—as necessary—this is a sense that's no longer respectable. Mathematical proof is a rich tissue of ever-growing assumptions ingeniously grafted onto an already existing subject matter; and where proof is a combination of steps taken on the basis of such assumptions—tacitly or explicitly—as well as on the basis of purely combinatorial inferences that are recognizably valid only because after repeated checking we can see that no mistake has been made in the complex and intricate manipulations, and not because of the sense of valid inference that carries us along as we calculate. Still left, of course, is the intuitive sense of a guarantee that the contemplation of examples of classical inferences—like modus ponens—gives us.

One point that Etchemendy (1990, 2) may be taken to be indicating when he writes "any sentence is derivable from any other in *some* such system" is that syntactic rules can be so arbitrary in their quality that, given any two sentences *A* and *B*, a set of rules can be found so that *B* is derivable from *A*. We (intuitively) recognize this about syntactic rules, and certainly when we contemplate the rule of modus ponens, our perception of it isn't restricted to the impression that it's a (syntactic) rule that we happen to follow. There is something compelling about it which is not shared with all the arbitrarily different syntactic rules we may imagine. This is why the discovery of alternative logics—effective ones—is the discovery of many inference systems most of which, in some sense, we don't "understand": they don't seem "reasonable."

It's tempting, therefore, to go *metaphysical* at this point, to argue that we are sensing the contours of the metaphysically necessary—of how things *must* be. Why or how we could be endowed with such epistemic powers, of course, has been something philosophers have long failed to provide an explanation for. But the failure of an epistemological story for metaphysical insight is no objection to the claim that we have such metaphysical insight if we, on the other hand, can't supply another story about where this sense of necessity— one that settles upon the inferences of a classical logic (and not on its syntactically effective cousins)—comes from.

Quine's well-known story about our sense of the necessity of logical inference and truth draws on two resources. First, there is the centrality of logical principles. Logical principles are so ubiquitous that our ability to imagine how we could reason with alternatives fails us because of the magnitude of the imaginative task. Second, there is intellectual inertia: Long acquaintance and practice with a truth or inference gives us the impression that we understand it—and not its alternatives. Do these psychological facts suffice to explain the sense of a guarantee, that of course, this *must be*, when we contemplate an instance of modus ponens? Perhaps not. Some find the contemplation of a desertion of modus ponens simultaneously the contemplation of the desertion of *reason*: it's a desertion that *time* will not immure us to.

Still, the general Quinean strategy to look to aspects of our psychology to explain where this rigid sense of the rightness of modus ponens is coming from is the required strategy. There are aspects of our psychology, however, that Quine hasn't considered, and that can explain the intuition of the necessity of modus ponens. The crucial element—as chapter 9 indicated—is the interplay between those aspects of our concepts that we are aware of, and those equally crucial aspects of our concepts—which enable us to use and apply such concepts—that we are nonetheless unaware of.

Our conscious grasp of how we manipulate syntactic rules makes our abilities with them seem remarkably "plastic." Given an arbitrary algorithmic system, we can manipulate it at will (or so it seems). But, introspectively, when we reason, we don't feel as if we're manipulating syntactic rules. Instead, we seem to maneuver the steps—as mentioned—by a failure to see how it is possible for the later step to be false if the previous one is true. What we are conscious of when we reason gives us no perception of the possibility of an alternative, despite the fact that what's really going on is a kind of syntactic processing that, in fact, does allow alternatives. Indeed, one way to see how misleading this intuitive sense of a guarantee actually is, is to notice that— psychologically—the impression it gives us is as forceful when we're reasoning correctly as when we've made a mistake. The Cartesian explanation for why we make mistakes is that we skip steps. But this influential doctrine is belied by the fact that even when we engage in the explicit construction of derivations— where no steps are skipped—we make mistakes anyway. The Cartesian picture is just wrong: our intuitive sense of a guarantee in reasoning is due precisely to the fact that how we reason—both psychologically (in the sense of how we're actually carrying it out) and sociologically (in the sense of what rules we normatively take to govern reasoning) are inaccessible to introspection.

So our inability to contemplate how we could desert modus ponens and come to accept some other syntactic rule, despite the fact that our grasp of modus ponens is (ultimately) syntactic is this: We can't become introspectively aware of how it is that we grasp modus ponens, and it is this that makes it impossible for us to imagine how it would feel to deny the validity of that inference rule.

Here is a moral for the book as a whole: Once we resolve not to fall back on Platonism as an explanation of the nature of mathematical proof, all the strange things that are really going on behind mathematical proof can finally be seen: what sort of implicit scaffolding was in place to transform mathematics from an empirical subject matter (something, arguably, it was until the ancient Greeks took hold of it) into the utterly amazing and unique proof-driven subject it has been ever since.

General Conclusion

Twentieth-century Anglo-American philosophy began, as it has been famously put, with a linguistic turn—with the simultaneous concern, on the part of G. E. Moore and Bertrand Russell, for language. Moore's concern was with the natural languages we speak; Russell's, with formal languages. There arose almost immediately a tension between the apparent demands of the mother tongue and the needs of the sciences which seemed—especially in mathematics—to drive us beyond the bounds of that tongue.[1] I sometimes think the tension in question emerged (in the course of the last century) most clearly in two extreme poles of doing philosophy—the ordinary-language sort of approach exemplified by (many but not all) followers of Wittgenstein, and the regimentation style of philosophy exemplified by (many but not all) followers of Quine.

Illuminating mathematical practice has been at the center of the regimentation approach—for it was projects concerned with getting clear about the logic and epistemology of mathematics that motivated the study of formal languages to begin with. It's ironic, therefore, that one form of the maverick rebellion against regimentation was the simple observation that the resulting formalisms seemed wildly disconnected from both the ways that mathematicians

---

[1] One of my favorite examples of the snobbery that practitioners of "ordinary language philosophy" were capable of directing toward the warping of the mother tongue apparently required by science occurs when Austin (1962, 18), criticizing Ayer for his peculiar use of "perceiving indirectly," writes, "No doubt there are complications here (raised, perhaps, by the electron microscope, for example, about which I know little or nothing)." For contrast, see Hacking 1985.

reason—as, that is, they continue to reason in the vernacular—and the ways they come to know what they know.

I've tried to show our inescapable needs both to continue reasoning within the vernacular and to regiment ourselves out of that very vernacular while simultaneously still using it. Ordinary language isn't an ideal device for communication any more than the human back is an ideal device for standing upright—and for similar reasons: They are gerrymandered products of evolution. On the other hand, we are no more likely to desert human languages in the *near* future than we are to desert our own bodies.

Still there remains the question of how we successfully reason using human languages—languages that in so many ways are clearly not designed to optimally reflect good inferences. I've provided two specific illustrations of how we do it: The first was the culmination of part I, which described how we can continue to successfully reason in a logically inconsistent medium. The second was the topic of part II, which described how the proofs of mathematics, couched in the vernacular, indicate mechanically recognizable derivations. In both cases, the suggestion was made that our practice is in accord with (a theory of) a (largely nonexistent) structure that is optimally designed for successful inference, and that the good reasoners among us draw conclusions that correlate to what our inferences would yield if they were couched in that structure. In the process we (almost automatically) disregard aspects of ordinary language that otherwise block recognition of such inferences. Reasoning, thus, isn't easy—and perhaps isn't natural; and we're very much unequal in our abilities to exercise it. My story may go some way toward explaining such differences.

Mathematics, early on, drew the attention of philosophers because of its delightfully long chains of (nevertheless) successful inferences. I've largely, although not entirely, evaded the question of whether the first-order story I've told about mathematical reasoning extends beyond that subject matter to human reason generally. There is some evidence that it doesn't: What I called the "semantic (and pragmatic) excrescences" of the terms of ordinary language seem to come into their own even in the sciences—where, for example, notions of causation play havoc with the neat truth-functionality of "if—then." What I'd like to show, if I can see how, is that despite appearances, the first-order story I've told about mathematics does generalize. Mature mathematics isn't a specialized game of reason; its logic is the whole of our natural logic. But this is future work.

One last point: This qualification about the general applicability of the first-order descriptive thesis to our reasoning in general may seem to restrict my general theses to reasoning-in-mathematics, and thus it may seem that the title of this book is misleading. It isn't. Regardless of the fate of the descriptive first-order thesis for reasoning generally, it certainly remains the case that first-order reasoning is a part—if not a major part—of reasoning per se. And, in any case, the major claims about the opacity to introspection of the principles by which we reason remain intact without the first-order descriptive thesis—even more so.

Bibliography

Anderson, A. R. 1970. St. Paul's epistle to Titus. In *The paradox of the liar*, ed. R. L. Martin, 1–11. New Haven, Conn.: Yale University Press.

Anderson, Stephen R. 1988. Morphological change. In *Linguistic theory: Foundations,* ed. Frederick J. Newmeyer, 324–62. Vol. 1 of *Linguistics: The Cambridge survey.* Cambridge: Cambridge University Press.

Arbib, Michael A. 1990. *A Piagetian perspective on mathematical construction. Synthese* 84: 43–58.

Armour-Garb, B., and J. Beall. 2001. Can deflationists be dialetheists? *Journal of Philosophical Logic* 30(6): 593–608.

Austin, J. L. 1962. *Sense and sensibilia.* Oxford: Oxford University Press.

———. 1979. Truth. In *Philosophical papers*, ed. J. O. Urmson and G. J. Warnock, 117–33. 3rd ed. Oxford: Oxford University Press.

Azzouni, Jody. 1991. A simple axiomatizable theory of truth. *Notre Dame Journal of Formal Logic* 32(3): 458–93.

———. 1994. *Metaphysical myths, mathematical practice: The ontology and epistemology of the exact sciences.* Cambridge: Cambridge University Press.

———. 1995. Review of "The liar speaks the truth." *Mind* 104(413): 222–25.

———. 1997. Thick epistemic access: Distinguishing the mathematical from the empirical. *Journal of Philosophy* 94: 472–84.

———. 1998. On "on what there is." *Pacific Philosophical Quarterly* 79(1): 1–18.

———. 1999. Comments on Shapiro. *Journal of Philosophy* 96: 541–44.

———. 2000a. Applying mathematics: An attempt to design a philosophical problem. *Monist* 83: 209–27.

———. 2000b. *Knowledge and reference in empirical science.* London: Routledge. (Paperback edition [with important corrections] published in 2004.)

————. 2001. Truth via anaphorically unrestricted quantifiers. *Journal of Philosophical Logic* 30: 329–54.

————. 2004a. *Deflating existential consequence: A case for nominalism.* Oxford: Oxford University Press.

————. 2004b. Proof and ontology in mathematics. In *New trends in the history and philosophy of mathematics*, ed. Tinne Hoff Kjeldsen, Stig AndurPedersen, and Lise Mariane Sonne-Hansen, 117–33. Denmark: University Press of Southern Denmark.

————. 2004c. Theory, observation and scientific realism. *The British Journal for the Philosophy of Science* 55(3): 371–92.

————. 2005. Is there still a sense in which mathematics can have foundations? In *Essays on the foundations of mathematics and logic*, ed. Giandomenico Sica, 9–47. Milan, Italy: Polimetrica S.a.s.

Bar-On, Dorit, Claire Horisk, and William G. Lycan. 2000. Deflationism, meaning, and truth conditions. *Philosophical Studies* 101: 1–28.

Barwise, Jon. 1985. Chapter 1: Model-theoretic logics: Background and aims. In *Model-theoretic logics*, ed. J. Barwise and S. Feferman, 3–23. Berlin: Springer-Verlag.

————. 1989. Mathematical proofs of computer system correctness. *Notices of the American Mathematical Society* 36: 844–51.

Barwise, Jon, and Robin Cooper. 1981. Generalized quantifiers in natural language. *Linguistics and Philosophy* 4: 159–219.

Barwise, Jon, and Solomon Feferman. 1985. *Model-theoretic logics.* Berlin: Springer-Verlag.

Blackburn, Simon. 1984. *Spreading the word.* Oxford: Oxford University Press.

Bloor, David. 1983. *Wittgenstein: A social theory of knowledge.* New York: Columbia University Press.

Boolos, George. 1998a. A curious inference. In *Logic, logic, and logic*, ed. Richard Jeffrey, 376–82. Cambridge, Mass.: Harvard University Press.

————. 1998b. To be is to be a value of a variable (or to be some values of some variables). In *Logic, logic, and logic*, ed. Richard Jeffrey, 54–72. Cambridge, Mass.: Harvard University Press.

Brown, James Robert. 1999. *Philosophy of mathematics: An introduction to the world of proofs and pictures.* London: Routledge.

Burge, T. 1984. Semantic paradox. In *Recent essays on truth and the liar paradox*, ed. Robert L. Martin, 83–117. Oxford: Oxford University Press. First published in *Journal of Philosophy* 76 (1979): 169–78.

Burgess, John L. 1993. Review of Shapiro (1991). *Journal of Symbolic Logic* 58: 363–65.

Chihara, C. 1979. The semantic paradoxes: A diagnostic investigation. *Philosophical Review* 88: 590–618.

Chomsky, Noam. 1986. *Knowledge of language: Its nature, origin, and use.* New York: Praeger.

Corcoran, J. 1980. Categoricity. *History and Philosophy of Logic* 1: 187–207.

David, Marian. 1994. *Correspondence and disquotation.* Oxford: Oxford University Press.

Davidson, Donald. 1984a. In defense of convention T. In his *Inquiries into truth and interpretation*, 65–75. Oxford: Oxford University Press.

————. 1984b. On the very idea of a conceptual scheme. In his *Inquiries into truth and interpretation*, 183–98. Oxford: Oxford University Press.

————. 1984c. Semantics for natural languages. In his *Inquiries into truth and interpretation*, 55–64. Oxford: Oxford University Press.

————. 1984d. Theories of meaning and learnable languages. In his *Inquiries into truth and interpretation*, 3–15. Oxford: Oxford University Press.

————. 1984e. Truth and meaning. In his *Inquiries into truth and interpretation*, 17–36. Oxford: Oxford University Press.

————. 1996. The folly of trying to define true. *The Journal of Philosophy* 93, 263–78.

Dehaene, S. 1997. *The number sense*. Oxford: Oxford University Press.

DeMillo, Richard A., Richard J. Lipton, and Alan J. Perlis. 1979. Social processes and proofs of theorems and programs. *Communications of the ACM* 22, 271–80.

Descartes, R. 1931. Rules for the direction of the mind. In *The philosophical works of Descartes*, ed. and trans. Elizabeth S. Haldane and G. R. T. Ross, 1–77. London: Cambridge University Press.

Deutsch, Harry. 2000. Making up stories. In *Empty names, fiction, and the puzzles of non-existence*, ed. Anthony Everett and Thomas Hofwever, 149–81. Stanford, Calif.: CSLI Publications.

Devitt, Michael. 2002. The metaphysics of deflationary truth. In *What is truth?*, ed. Richard Schantz, 60–78. Berlin: Walter de Gruyter.

Dretske, Fred. 1981. *Knowledge and the flow of information*. Cambridge, Mass.: MIT Press.

Dudley, Underwood. 1987. *A budget of trisections*. New York: Springer-Verlag.

Dummett, Michael. 1959. Truth. In *Truth and other enigmas*, 1–24. Cambridge, Mass.: Harvard University Press.

————. 1999. Of what kind of thing is truth a property? In *Truth*, ed. S. Blackburn and K. Simmons, 264–81. Oxford: Oxford University Press.

Dunn, J. M., and Belnap, N. D., Jr. 1968. The substitution interpretation of the quantifiers. *Nous* 2: 177–85.

Etchemendy, John. 1990. *The concept of logical consequence*. Cambridge, Mass.: Harvard University Press.

Evans, Gareth. 1973. The causal theory of names. In *Collected papers*, 1–24. Oxford: Oxford University Press.

Felsager, B. 1998. *Geometry, particles, and fields*. New York: Springer-Verlag.

Feferman, S. 1982. Towards useful type-free theories, I. *Journal of Symbolic Logic* 49(1): 75–111.

Fetzer, James H. 1988. Program verification: The very idea. *Communications of the ACM* 31: 1048–63.

Field, Hartry. 1989. Introduction: Fictionalism, epistemology, and modality. In *Realism, mathematics, and modality*, 1–52. Oxford: Basil Blackwell.

————. 2001a. Deflationist views of meaning and content. In *Truth and the absence of fact*, 104–40. Oxford: Oxford University Press.

————. 2001b. Postscript: Deflationist views of meaning and content. In *Truth and the absence of fact*, 141–56. Oxford: Oxford University Press.

————. 2001c. Tarski's theory of truth. In *Truth and the absence of fact*, 3–26. Oxford: Oxford University Press.

Flum, J. 1985. Chapter 3: Characterizing logics. In Model-theoretic logics, ed. J. Barwise and S. Feferman, 77–120. Berlin: Springer-Verlag.

Fodor, Jerry A. 1975. *The language of thought*. Cambridge, Mass.: Harvard University Press.

————. 1990. A theory of content (I and II), in *A theory of content and other essays*, 51–136. Cambridge, Mass.: MIT Press.

Gould, S. J., and R. C. Lewontin. 1978. The spandrels of San Marco and the Pan-glossian paradigm: A critique of the adaptionist programme, Proceedings of the Royal Society of London 205: 581–98.

Grover, D. L., and N. D. Belnap. 1992. Quantifying in and out of quotes. In *A prosentential theory of truth*, 244–75. Princeton, N.J.: Princeton University Press.

Grover, D. L., J. L. Camp, and N. D. Belnap. 1992. A prosentential theory of truth. In *A prosentential theory of truth*, 70–120. Princeton, N.J.: Princeton University Press.

Grover, D. L. 2002. On locating our interest in truth. In *What is truth?*, ed. Richard Schantz, 120–32. Berlin: Walter de Gruyter.

Gupta, Anil. 1984. Truth and paradox. In *Recent essays on truth and the liar paradox*, ed. Robert L. Martin, 15–235. Oxford: Oxford University Press. First published in *Journal of Philosophical Logic* 11 (1982): 1–60.

———. 1999. A critique of deflationism. In *Truth*, ed. S. Blackburn and K. Simmons, 282–307. Oxford: Oxford University Press.

———. 2002. An argument against Tarski's convention T. In *What is truth?*, ed. Richard Schantz, 225–37. Berlin: Walter de Gruyter.

Hacking, Ian. 1985. Do we see through a microscope? In *Images of science: Essays on realism and empiricism, with a reply from Bas C. van Fraassen*, ed. Paul M. Churchland and C. A. Hooker, 132–52. Chicago: University of Chicago Press.

Halbach, V. 1999. Disquotationalism and infinite conjunctions. *Mind* 108(411): 249–85.

Heck, R. G., Jr. 1997. Tarski, truth, and semantics, *Philosophical Review* 106(4): 533–54.

Hempel, Carl G. 1965. A logical appraisal of operationism. In *Aspects of scientific explanation and other essays in the philosophy of science*, 123–33. New York: Free Press.

Hersh, Ruben. 1997. *What is mathematics, really?* Oxford: Oxford University Press.

Herstein, I. N. 1975. *Topics in algebra*. 2nd ed. Lexington, Mass.: Xerox College Publishing.

Herzberger, Hans. 1964. The logical consistency of language. *Harvard Educational Review* 35: 469–80.

———. 1966. The logical consistency of language. In *Language and learning (A revision and expansion of the 1964 special issue of the Harvard Educational Review)*, 250–63. New York: Harcourt, Brace.

———. 1967. The truth-conditional consistency of natural languages. *Journal of Philosophy* 64: 29–35.

———. 1970. Paradoxes of grounding in semantics. *Journal of Philosophy* 67: 145–67.

———. 1982. Notes on naïve semantics. *Journal of Philosophical Logic* 11: 61–102. Reprinted in *Recent essays on truth and the liar paradox*, ed. Robert L. Martin, 133–74. Oxford: Oxford University Press, 1984.

Herzberger, Hans, and J. J. Katz. 1967. The concept of truth in natural languages. Typescript.

Hilbert, David. *Foundations of geometry*. Translated by Leo Unger. LaSalle, Ill.: Open Court.

Hoddeson, L., L. Brown, M. Riordan, and M. Dresden. 1997. *The rise of the standard model*. Cambridge: Cambridge University Press.

Hodges, Wilfred. 1983. Elementary predicate logic. In *Handbook of philosophical logic*, ed. D. Gabbay and F. Guenthner, I: 1–131. Dordrecht: D. Reidel.

Hofweber, Thomas. 2005. Inexpressible properties and propositions. In *Oxford studies in metaphysics,* vol. 2, ed. Dean Zimmerman. Oxford: Oxford University Press.

Horwich, Paul. 1995. Disquotation and cause in the theory of reference. In *Philosophical Issues,* vol. 6, ed. Enrique Villanueva, 73–78. Atascadero, Calif.: Ridgeview.

————. 1997. Deflationary truth and the problem of aboutness. In *Philosophical Issues*, vol. 8, ed. Enrique Villanueva, 95–106. Atascadero, Calif.: Ridgeview.

————. 1998. *Truth*. 2nd ed. Oxford: Oxford University Press.

————. 1999. The minimalist conception of truth. In *Truth*, ed. S. Blackburn and K. Simmons, 239–63. Oxford: Oxford University Press.

Jesseph, Douglas M. 1999. *Squaring the circle: The war between Hobbes and Wallis*. Chicago: University of Chicago Press.

Kant, Immanuel. 1998. *Critique of pure reason*. Trans. and ed. Paul Guyer and Allen W. Wood. Cambridge: Cambridge University Press.

Katz, Jerrold J. 1972. *Semantic theory*. New York: Harper & Row.

————. 2004. *Sense, reference, and philosophy*. Oxford: Oxford University Press.

Keenan, E. L., and D. Westerståhl. 1997. Generalized quantifiers in linguistics and logic. In *Handbook of logic and language*, ed. Johan van Benthem and Alice ter Meulen, 837–93. Cambridge, Mass.: MIT Press.

Ketland, J. 1999. Deflationism and Tarski's paradise. *Mind* 108(429): 69–94.

Kirkham, R. L. 1992. *Theories of truth*. Cambridge, Mass.: MIT Press.

Kleene, S. C. 1971. *Introduction to metamathematics*. Amsterdam: North-Holland.

Kline, Morris. 1980. *Mathematics: The loss of certainty*. Oxford: Oxford University Press.

Kneale, William, and Martha Kneale. 1984. *The development of logic*. Oxford: Oxford University Press.

Kreisel, Georg. 1967. Informal rigour and completeness proofs. In *Problems in the philosophy of mathematics*, ed. I. Lakatos, 138–86. Amsterdam: North-Holland.

————. 1970. Principles of proof and ordinals implicit in given concepts. In *Intuitionism and proof theory. Proceedings of the summer conference at Buffalo, N.Y., 1968*, ed. A. Kino, J. Myhill, and R. E. Vesley, 489–516. Amsterdam: North-Holland.

Kripke, Saul. 1976. Is there a problem about substitutional quantification? In *Truth and meaning: Essays in semantics*, ed. G. Evans and J. McDowell, 325–419. Oxford: Oxford University Press.

————. 1982. *Wittgenstein on rules and private language*. Cambridge, Mass.: Harvard University Press.

————. 1984. Outline of a theory of truth. In *Recent essays on truth and the liar paradox*, ed. Robert L. Martin, 53–81. Oxford: Oxford University Press. First published in *Journal of Philosophy* 72 (1975): 690–716.

Lakatos, Imre. 1976. *Proofs and refutations: The logic of mathematical discovery*. Cambridge: Cambridge University Press.

Lance, Mark. 1997. The significance of anaphoric theories of truth and reference. *Philosophical Issues* 8, ed. Enrique Villanueva, 181–98. Atascadero, Calif.: Ridgeview.

Lavine, Shaughan. 1994. *Understanding the infinite*. Cambridge, Mass.: Harvard University Press.

Leblanc, Hugues. 1983. Alternatives to standard first-order semantics. In *Handbook of Philosophical Logic*, ed. D. Gabbay and F. Guenthner, I: 189–274. Dordrecht: D. Reidel.

Leeds, Stephen. 1978. Theories of reference and truth. *Erkenntnis* 13: 111–29.

Lindström, Per. 1969. On extensions of elementary logic. *Theoria* 35: 1–11.

Longstaff, Ben. 2003. Interview with Marcus du Sautoy. *New Scientist* (Nov. 22–28, 2003): 48–51.

Lynch, Michael P. 2000. Alethic pluralism and the functionalist theory of truth. *Acta Analytica* 15(24): 195–214.

MacKenzie, Donald. 2001. *Mechanizing proof: Computing, risk, and trust.* Cambridge, Mass.: MIT Press.

Makowsky, J. A. 1985. Chapter 20: Abstract embedding relations. In *Model-theoretic logics*, ed. J. Barwise and S. Feferman, 747–91. Berlin: Springer-Verlag.

Martin, R. L. 1970. A category solution to the liar. In *The paradox of the liar*, ed. Robert L. Martin, 91–120. New Haven, Conn.: Yale University Press.

McGee, Vann. 1992. Maximal consistent sets of instances of Tarski's schema (T). *Journal of Philosophical Logic* 21: 235–41.

———. 1993. A semantic conception of truth? *Philosophical Topics* 21: 83–111.

Moore, G. H. 1980. Beyond first-order logic: The historical interplay between mathematical logic and axiomatic set theory. *History and Philosophy of Logic* I: 95–137.

———. 1982. *Zermelo's axiom of choice: Its origins, development, and influence.* Berlin: Springer-Verlag.

Orwell, George. 1949. *1984.* New York: Harcourt, Brace.

Parsons, Charles. 1984. The liar paradox. In *Recent essays on truth and the liar paradox*, ed. Robert L. Martin, 9–45. Oxford: Oxford University Press. First published in *Journal of Philosophical Logic* 3 (1974): 381–412.

Parsons, Terence. 1980. *Nonexistent objects.* New Haven, Conn.: Yale University Press.

Patterson, Douglas. Forthcoming (a). Deflationism and the truth conditional theory of meaning. *Philosophical Studies.*

———. Forthcoming (b). Tarski on the necessity reading of convention T. *Synthese.*

Priest, Graham. 1987. *In contradiction.* Dordrecht: Martinus Nijhoff.

Putnam, Hilary. 1975a. Do true assertions correspond to reality? In *Mind, language, and reality: Philosophical papers*, vol. 2, 70–84. Cambridge: Cambridge University Press.

———. 1975b. The meaning of 'meaning'. In *Mind, language and reality: Philosophical papers*, vol. 2, 139–52. Cambridge: Cambridge University Press.

———. 1978. Meaning and knowledge. In *Meaning and the moral sciences*, 1–80. London: Routledge and Kegan Paul.

———. 1983. Models and reality. In his *Philosophical papers*, vol. 3, 1–25. Cambridge: Cambridge University Press.

Quine, W. V. 1960. *Word and object.* Cambridge, Mass.: MIT Press.

———. 1970. *Philosophy of Logic.* Englewood Cliffs, N.J.: Prentice Hall.

———. 1975a. Quantifiers and propositional attitudes. In *The ways of paradox and other essays*, 185–96. 2nd ed. Cambridge, Mass.: Harvard University Press.

———.1975b. Three grades of modal involvement. In *The ways of paradox and other essays*, 158–76. 2nd ed. Cambridge, Mass.: Harvard University Press.

———. 1975c. Truth by convention. In *The ways of paradox and other essays.* 77–106. 2nd ed. Cambridge, Mass.: Harvard University Press.

———. 1980a. Logic and the reification of universals. In *From a logical point of view*, 102–29. 2nd ed. Cambridge, Mass.: Harvard University Press.

———. 1980b. Reference and modality. In *From a logical point of view*, 139–59. 2nd ed. Cambridge, Mass.: Harvard University Press.

———. 1986. Reply to David Kaplan. In *The Philosophy of W. V. Quine*, ed. Lewis Edwin Hahn and Paul Arthur Schilpp, 290–94. La Salle, Ill.: Open Court.

Quine, W. V., and Goodman, Nelson. 1972. Steps towards a constructive nominalism. In Nelson Goodman, *Problems and projects*, 173–98. New York: Bobbs-Merrill. First published in *Journal of Symbolic Logic* 12 (1947): 105–22.

Rav, Yehuda. 1999. Why do we prove theorems? *Philosophia Mathematica* (3) 7: 5–41.

Resnik, Michael D. 1997. *Mathematics as a science of patterns*. Oxford: Oxford University Press.

Richard, Mark. 1996. Propositional quantification. In *Logic and reality: Essays on the legacy of Arthur Prior*, ed. B. J. Copeland, 437–60. Oxford: Oxford University Press.

Rogers, Hartry, Jr. 1967. *Theory of recursive functions and effective computability*. New York: McGraw-Hill.

Rorty, Richard. 1991. *Objectivity, relativism, and truth: Philosophical papers*, vol. 1. Cambridge: Cambridge University Press.

Routley, Richard. 1980. *Exploring Meinong's jungle and beyond*. Canberra, Australia: Departmental Monograph #3, Philosophy Department, Research School of Social Sciences, Australian National University.

Shapiro, S. 1991. *Foundations without foundationalism: A case for second-order logic*. Oxford: Oxford University Press.

———. 1998. Proof and truth: Through thick and thin. *Journal of Philosophy* 95(10): 493–521.

———. 1999. Do not claim too much: Second-order logic and first-order logic. *Philosophia Mathematica* (3) 7: 42–64.

Sher, Gila. 2004. In search of a substantive theory of truth. *Journal of Philosophy* CI(1): 5–36.

Simmons, K. 1999. Deflationary truth and the liar. *Journal of Philosophical Logic* 28(5): 455–88.

Soames, Scott. 1997. The truth about deflationism. *Philosophical Issues* 8, ed. Enrique Villanueva, 1–44. Atascadero, Calif.: Ridgewood.

———. 1999. *Understanding truth*. Oxford: Oxford University Press.

Stampe, Dennis. 1977. Towards a causal theory of linguistic representations. In *Midwest studies in philosophy 2*, ed. P. French, T. Euhling, and H. Wettstein, 42–63. Minneapolis: University of Minneapolis Press.

Steinbring, Heinz. 1991. The concept of chance in everyday teaching: Aspects of social epistemology of mathematical knowledge. *Educational Studies in Mathematics* 22: 502–22.

Strawson, P. F. 1999. Truth. In *Truth*, ed. S. Blackburn and K. Simmons, 162–82. Oxford: Oxford University Press.

Tarski, Alfred. 1944. The semantic conception of truth. *Philosophy and Phenomenological Research* 4(3): 341–75.

———. 1983a. The concept of truth in formalized languages. In *Logic, semantics, metamathematics*, ed. J. Corcoran, 152–278. Indianapolis, Ind.: Hackett.

———. 1983b. On the concept of logical consequence. In *Logic, semantics, metamathematics*, ed. J. Corcoran, 409–20. Indianapolis, Ind.: Hackett.

van Inwagen, Peter. 1981. Why I don't understand substitutional quantification. *Philosophical Studies* 39: 281–85.

———. 2000. Quantification and fictional discourse. In *Empty names, fiction, and the puzzles of non-existence*, ed. Anthony Everet and Thomas Hofweber, 235–47. Stanford, Calif.: CSLI Publications.

Varzi, Achille C. 2002. Worlds and objects. In *Individuals, essence, and identity. Themes of analytic metaphysics*, ed. A. Bottani, M. Carrara, and D. Giaretta, 49–75. Dordrecht: Kluwer.

Westfall, Richard S. 1980. *Never at rest: A biography of Isaac Newton*. Cambridge: Cambridge University Press.

Williams, Michael. 1988. Epistemological realism and the basis of scepticism. *Mind* 97: 415–39.

————. 2002. On some critics of deflationism. In *What is truth?*, ed. Richard Schantz, 146–58. Berlin: Walter de Gruyter.

Wittgenstein, Ludwig. 1953. *Philosophical investigations.* Trans. G. E. M. Anscombe. New York: Macmillan.

Wright, Crispen. 1992. *Truth and objectivity.* Cambridge, Mass.: Harvard University Press.

Index

abstract completeness. *See* recursive enumerability

abstracta, as socially constructed, 130. *See also* Platonism

ADs. *See* arguments that indicate derivations

Albert, David, 132n16

algorithmic systems, 119, 121, 137n33, 143–44, 147–48, 152–53n34, 153–54, 162–67, 174–75, 176n15, 187–88, 197, 198, 205–6

ambiguity, how truth conditions can handle, 28, 52–54, 69–70

anaphora, 22–23, 57–58, 62, 74, 79, 113n8. *See also* anaphorically unrestricted pronouns

anaphorically unrestricted pronouns, 23–25, 30, 35, 60, 215

Anaphorish, 23, 24–25, 28, 29, 30, 32n34, 199

Anderson, A. R., 81

Anderson, S. R., 136n31

Arbib, M. A., 169

argument-schemata, 212–13, 215n38

arguments that indicate derivations, 149–50

Aristotle, 211n30, 215n38

Armour-Garb, B., 12, 74

AU-satisfaction, definition of, 63

Austin, J. L., 15, 16n8, 229n1

axiom of choice, the, 159n46

Ayer, A. J., 229n1

Bar-On, D., 25–26

Barwise, J., 145n15, 149, 192n1, 194–95, 196

Beall, J., 12, 74

Belnap, N., 58n44, 65n9

benign fixation of mathematical practice, the, 144, 225
attempted explanations of, 129–30
definition of, 129
derivations underlying, 142
how contemporary mathematics enables, 137–38
role of assumptions about mathematical abstracta in, 135–36
role of domain of application in, 133